T0302205

Synthesis of Pyrrol-based Bioconjugates

In recent years, the use of bioconjugation as strategy for development of more specific and directed drugs has been widely researched. Tetrapyrrolic macrocycles are one of the most common families of organic compounds applied in medicinal chemistry, specifically for diagnosis and therapy. For optimization of their efficiency as therapeutic molecules, it is relevant to promote their linking to biomolecules capable of targeting altered cells. This work brings a new dimension to the literature by combining structural modification of biomolecules and polymers with pyrrole-based compounds. It will be of great interest to academicians, industrialists, and undergraduate and graduate students of chemistry, biosciences, and pharmacy since, apart from the theoretical aspects, it discusses experimental components.

Features:

- Discusses methods for protein modification.
- Useful for academicians, industrialists, and undergraduate and graduate students of chemistry, biosciences, and pharmacy.
- Explains the synthesis of pyrrole-based molecules.

Synthesis of Pyrrol-based Bioconjugates

Perspectives and Applications

Mariette M. Pereira, Sara M.A. Pinto,
and Lucas D. Dias

CRC Press is an imprint of the
Taylor & Francis Group, an **informa** business

Front cover image: Juan Gaertner/Shutterstock

First edition published 2024
by CRC Press
2385 NW Executive Center Drive, Suite 320, Boca Raton FL 33431

and by CRC Press
4 Park Square, Milton Park, Abingdon, Oxon, OX14 4RN

CRC Press is an imprint of Taylor & Francis Group, LLC

ISBN: 978-0-367-63457-5 (hbk)
ISBN: 978-0-367-63458-2 (pbk)
ISBN: 978-1-003-11926-5 (ebk)

DOI: 10.1201/9781003119265

Typeset in Minion
by Apex CoVantage, LLC

Contents

About the Authors

Mariette M. Pereira is currently a full professor in organic chemistry at the University of Coimbra (Portugal). She obtained her PhD in organic chemistry in 1992 at the University of Coimbra and worked as a fellow assistant at the Universities of Liverpool in 1993 and Autónoma de Barcelona in 1998. She was the director of the chemistry research laboratory of the pharmaceutical *spin-off* Luzitin, SA, until 2015. Her research interests are focused on the synthesis of phosphorous-based chiral ligands and metal complexes for the development of catalysts for carbonylation tandem reactions. She is also strongly engaged in the synthesis of tetrapyrrolic macrocycle–based photosensitizers for applications on photodynamic therapy of cancer and photoinactivation of microorganisms. She published *ca.* 200 peer-reviewed papers, 20 books/book chapters, and she is the inventor of six WO patents. She won the Prix Tremplin Mariano Gago 2022 from the Portugal and Paris Academy of Science and the Hungary Academy of Science Distinguished Guest Scientist fellowship, 2023.

Sara M.A. Pinto is currently an auxiliary researcher at the University of Coimbra (Portugal). She obtained her PhD in macromolecular chemistry in 2012 at the University of Coimbra and began postdoctorate study at the University of Coimbra, with the subject of design and development of new multimodal contrast agents for medical imaging. Her current research interests are the design of tetrapyrrolic macrocycles for medical imaging applications, with particular emphasis on theranostics for new contrast agents for MRI/FI or MRI/PET. She has published 35 peer-reviewed papers, is co-inventor of two provisional patents, and is author of more than 50 communications (oral and poster) in national/international congresses.

Lucas D. Dias received his pharmacy diploma from the State University of Goiás (Brazil) and his PhD in chemistry (*catalysis and sustainability—CATSUS*) at the University of Coimbra, supervised by Prof. Mariette M. Pereira. Then, from 2019 to 2022, he performed as a postdoctoral research intern at the University of São Paulo (USP), Brazil, supervised by Prof. Vanderlei Salvador Bagnato. Currently, he has been an assistant professor at the Universidade Evangélica de Goiás (Brazil) and a researcher at the Coimbra Chemistry Centre (CQC) (University of Coimbra, Portugal). His current research interests are the design and synthesis of photocatalysts, homogeneous/heterogeneous catalysts based on tetrapyrrolic macrocycles for the activation of small molecules (O_2, CO, and CO_2), mechanisms in photodynamic therapy, and synthesis of photosensitizers for light-based reactions applications. He has published 50 peer-reviewed papers in national/international journals, is the author of more than 50 communications (oral and posters) in national/international scientific meetings, and is the inventor of three patents.

Preface

THE TEACHING TEAM OF the *Synthesis of Bioconjugates* course, included in the medicinal chemistry undergraduate program at the University of Coimbra, recognized the importance of providing a comprehensive support material to their students during online classes. As a result, they embarked on the writing of this book. The exchange of ideas between teachers and students, coupled with the course's problem-solving provided during the pandemic period, ultimately led to the creation of the final text. This book is the product of a collaborative effort between the students and teachers and serves as a valuable resource for anyone interested in the field of bioconjugate synthesis.

Over the course of the following two years, from 2021 to 2023, we gradually started face-to-face classes and explored the topic in more detail. Through our studies, we are continuously recognizing the vast potential of porphyrins as a prime example of synthetic molecules that can be easily conjugated to biomolecules. Furthermore, we discovered that bioconjugate derivatives of porphyrins hold tremendous promise for a wide range of medicinal applications, including therapeutic and immunotherapeutic uses, as well as in clinical imaging. The knowledge and insights gained from our face-to-face classes have significantly influenced the content of this book. As a result, readers can expect comprehensive coverage of porphyrin-based bioconjugates that reflect the latest understanding and research in this exciting field.

It is noteworthy that, despite the difficult circumstances that both students and teachers have faced during the past three years of the pandemic, this support text has proven to be a critical factor in sparking students' interest in the *Synthesis of Bioconjugates* course and related topics. As a result, we have observed exceptional final results among our students, with approval rates exceeding 95%. This outcome reinforces the significance of providing comprehensive and captivating learning materials, particularly during times of crisis. Such resources can help students maintain their motivation and achieve their academic goals, even in the face of challenging circumstances. The success of our students is a good evidence of the importance of investing in quality educational resources that support and engage learners.

The *Synthesis of Bioconjugates* course was designed to be centered on the needs of the students, utilizing a continuous evaluation model that emphasized research assignments focused on bioconjugates and their clinical applications. The support text was instrumental in providing a solid foundation and motivating students to explore the diverse topics in depth.

To evaluate the students' understanding, we incorporated oral presentations and lively discussions with their peers and the teaching team, which accounted for 60% of the overall grade. In addition, we conducted a comprehensive written test that represented 40% of the final grade. This approach ensured that students had a thorough understanding of the course material and were well-prepared to apply their knowledge in practical settings.

Overall, our approach to course design and evaluation fostered a dynamic and engaging learning environment that empowered students to take ownership of their education and achieve their full potential. It was written aiming to serve as an excellent resource for students studying medicinal chemistry, medicine, biochemistry, and pharmacy. The book is also highly relevant for researchers working in the pharmaceutical industry who are interested in exploring new applications of porphyrins in clinical settings. The team is confident that this textbook will be a valuable addition to the scientific literature and contribute to furthering the knowledge of bioconjugates and their clinical relevance.

This book is divided into six chapters. In the first chapter, we give a brief introduction to the basic concepts related to the synthesis of bioconjugates, including reagents and synthetic strategies. Relevant reviews and textbooks are mentioned, for complementary reading.

The second chapter provides an overview of the main techniques used for the synthesis of porphyrins and highlights key considerations for functionalizing these molecules. Additionally, the most relevant mechanisms of the reactions involved are presented in detail. The chapter offers an in-depth analysis of these topics, which helps readers build a strong understanding of the subsequent chapters that focus on the synthesis and applications of porphyrin-based bioconjugates.

In the third chapter, the different strategies for the synthesis of dendrimers are presented, including examples of polyester-, polyamidoamine-, and poly(peptide)-type dendrimers. Selected examples of synthesis and applications of dendrimer–porphyrin conjugates are also presented and discussed.

Implementation of PEGylation reactions are described in Chapter 4. Selected examples of preparation and applications of PEG-porphyrin-based conjugates are discussed and described.

Chapter 5 describes selected methods for bioconjugation of functionalized porphyrins with carbohydrates, including detailed reaction mechanisms. Several examples of porphyrin-based carbohydrate bioconjugates are also presented in this chapter, including detailed mechanisms.

Finally, Chapter 6 analyzes selected methods to prepare peptide–porphyrin bioconjugates. Detailed mechanisms are also included.

We would like to express our gratitude to all the students involved in this teaching project, particularly those from the academic year 2022–2023, who contributed to the final proofreading (see **Figure 0.1**).

We would also like to thank our colleagues from the Catalysis and Fine Chemistry group at the Coimbra Chemistry Centre, namely, Rafael Aroso,

FIGURE 0.1 Class of students who attended.

Source: Synthesis of Bioconjugates course.

Mário Calvete, Rui Carrilho, and Fábio Rodrigues, for their proofreading assistance.

We are also grateful to the publisher Taylor & Francis, particularly Sukirti Singh and Hillary Lafoe, who believed in this work and provided us with great encouragement to complete it.

Mariette M. Pereira
Sara M.A. Pinto
Lucas D. Dias

CHAPTER 1

Introduction

Bioconjugates are characterized by the linkage of two or more molecules wherein at least one molecule is of biological origin and the other(s) is (are) synthetic, resulting in a single entity. The process of bioconjugation primarily entails the covalent attachment of one molecule to another. These chemical entities can either be directly linked or connected through a spacer fragment (**Figure 1.1**). This spacer molecule serves the purpose of enabling covalent binding between the two molecules, endowing flexibility to the final bioconjugate, while preserving the unique characteristics of each individual molecule [1,2].

When devising the synthesis of any bioconjugate, specific reagents may be required to modify the existing functional groups present in the individual molecules, enabling their coupling. For instance, the direct reaction between the thiol group of cysteine and the amine group of any molecule is not feasible. However, by prior transformation of the amine group into a carboxylic group using an anhydride, the bioconjugation becomes viable. It is important to note that, when working with biological molecules, the use of conventional organic reaction conditions is often impractical. These typically involve harsh conditions and toxic reagents, occasionally yielding unwanted side products. As an illustrating example, if the goal is to connect a peptide to a molecule containing an NH_2 residual group via a -COOH group, an advisable strategy is to employ an activating agent like *N,N′*-dicyclohexylcarbodiimide (DCC) or 1-ethyl-3-(3-dimethylaminopropyl)carbodiimide (EDC), among others. This approach ensures efficient and selective bioconjugation while minimizing by-products [1,2].

DOI: 10.1201/9781003119265-1

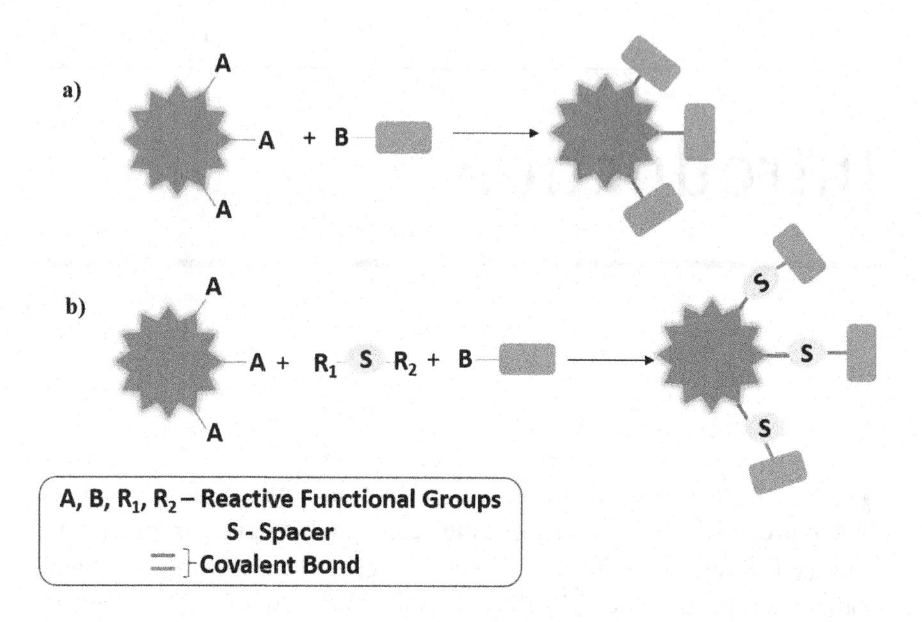

FIGURE 1.1 General structure of a bioconjugate molecule obtained via (a) direct covalent linking, (b) through a spacer.

Another crucial aspect when considering the synthesis of bioconjugates is the concept of protecting and deprotecting functional groups. Indeed, biological and synthetic molecules may possess several functional groups with similar reactivities, which can lead to non-selective binding. To overcome this drawback, selective protection strategies are employed to temporarily shield specific functional groups (FG). These protective groups (PG) ensure the desired selectivity during the bioconjugation process. Subsequently, these groups are selectively removed (deprotected) to reveal the functional groups necessary for the intended binding, allowing for controlled and targeted conjugation (**Figure 1.2**). An effective PG should possess the following key properties: (a) allow a selective and straightforward introduction to the desired FG; (b) show stability against the reaction conditions employed in subsequent reaction steps; (c) permit selective removal under moderate conditions when deprotection is necessary. Given that protection and deprotection strategies for functional groups have been extensively covered in other resources [3,4,5,6], this book will not delve into a comprehensive review or discussion of these techniques. However, it is important to highlight the importance of employing suitable protective groups to achieve the desired outcomes in bioconjugate

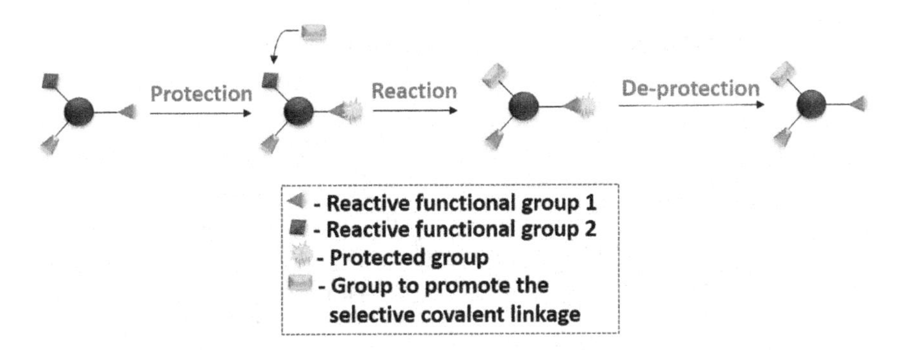

FIGURE 1.2 Schematic illustration of protection/deprotection strategy.

synthesis. In the subsequent chapters, whenever the utilization of specific protection and deprotection strategies is necessary, these will be thoroughly discussed, accompanied by detailed explanations of the involved mechanisms.

As mentioned previously, the formation of bioconjugates often involves the incorporation of a spacer fragment. Cross-linking reagents are commonly employed for this purpose and can be classified as either homobifunctional, possessing the same reactive groups at opposite ends of the cross-linker structure, or heterobifunctional, featuring different reactive groups at each end (Figure 1.3). The use of such reagents offers notable advantages. Firstly, it reduces the number of synthetic steps required to obtain the bioconjugate. Secondly, it enables precise control over the spatial distance between the involved molecules. This control proves advantageous as it mitigates issues like steric hindrance or unfavorable interactions that may arise between the biological and synthetic components [1,2].

With the continuous advances in science, novel synthetic approaches for the synthesis of bioconjugates have emerged, allowing structural modifications or binding processes to occur within biological environments, without interfering with ongoing biochemical processes. These innovative strategies have paved the way for more targeted and biocompatible bioconjugate synthesis. The chemistry employed in such modifications is known as biorthogonal chemistry, which offers enhanced selectivity and minimizes the formation of by-products. Briefly, the biorthogonal process involves two key steps: (a) incorporation of the biorthogonal group into the biomolecule and (b) reaction between the compound carrying an appropriate functional group that selectively reacts with the biorthogonal group, resulting in the desired bioconjugate. This biorthogonal approach

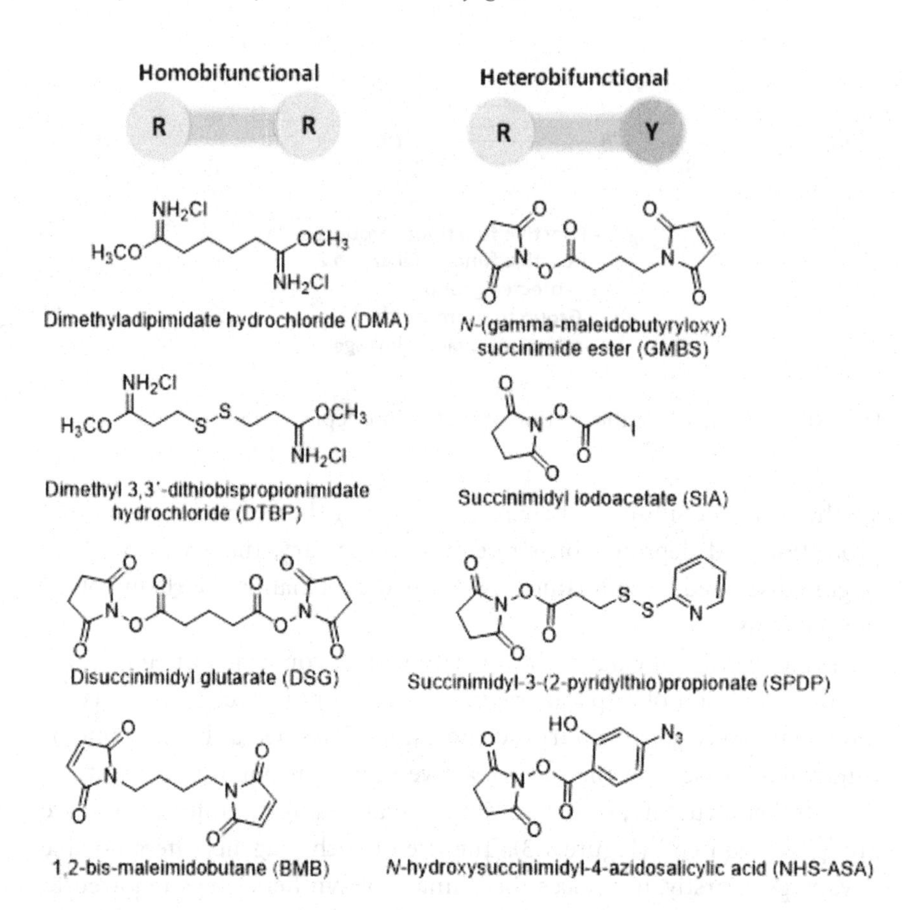

FIGURE 1.3 Examples of homobifunctional and heterofunctional reagents.

enables precise and controlled bioconjugation while minimizing undesired reactivity. It is important to emphasize that, for a reaction to be classified as biorthogonal, it must meet specific criteria. These include the ability to occur at physiological pH and temperatures, as well as to provide highly stable products that remain unaffected by water or other reactive biological residues. By meeting these requirements, biorthogonal reactions ensure compatibility with biological systems and offer robust and reliable bioconjugation processes [1,2,7]. The remarkable novelty, applicability, and benefits associated with this chemistry field were acknowledged in 2022, when the Nobel Prize in Chemistry was awarded to Sharpless, Bertozzi, and Meldal. Their groundbreaking contributions in the field have revolutionized bioconjugation and have had a profound impact on diverse scientific disciplines [8,9,10].

The significant potential of employing bioconjugates in medicine has been the subject of extensive research, with numerous examples documented in the literature. An illustrative relevant example consists of attaching a drug molecule to a biological carrier, such as a protein or antibody. This bioconjugate would enable precise targeting of the drug to specific cells or tissues within the body, thereby enhancing its efficacy while minimizing potential side effects. Such advancements in bioconjugate technology hold great promise for improving therapeutic outcomes in medical applications [11,12]. Furthermore, considerable efforts have led to the development of many bioconjugates specifically designed to target cancer cells. These tailored bioconjugates offer a more targeted and efficient approach to chemotherapy, resulting in enhanced efficacy while concurrently reducing adverse side effects [13,14].

Another relevant application of bioconjugates relates to diagnostic imaging. For example, by coupling a radiolabeled or other imaging agent with a biological molecule, such as an antibody or protein, it becomes possible to visualize and target specific cells or tissues within the body. This enables precise and non-invasive imaging, facilitating the detection and diagnosis of various diseases or medical conditions. The utilization of bioconjugates in diagnostic imaging holds immense potential for advancing medical diagnostics and enhancing patient care [15,16]. In addition, bioconjugates also allow the development of biomarkers for detecting *in vivo* diseases and the creation of novel therapies for autoimmune diseases (**Figure 1.4**) [17,18].

Despite the significant advantages of using bioconjugates, there remain critical aspects that require further attention. One such aspect is the challenging and time-consuming nature of conjugating two or more molecules, demanding continuous research into new, highly efficient, and selective methods and strategies. Additionally, the development of advanced purification techniques is imperative to ensure the production of high-quality bioconjugates. This ongoing research aims to rationalize and enhance the bioconjugation process, ultimately expanding the scope of applications and maximizing the potential benefits of bioconjugate technology.

At this point, we should emphasize that the stability and shelf life of bioconjugates can still pose concerns, particularly when considering their clinical application. The stability issue becomes particularly relevant in clinical settings, where long-term storage and reliable performance are somewhat incompatible. Additionally, immunogenicity presents a challenge in the clinical use of bioconjugates. As bioconjugates consist of both

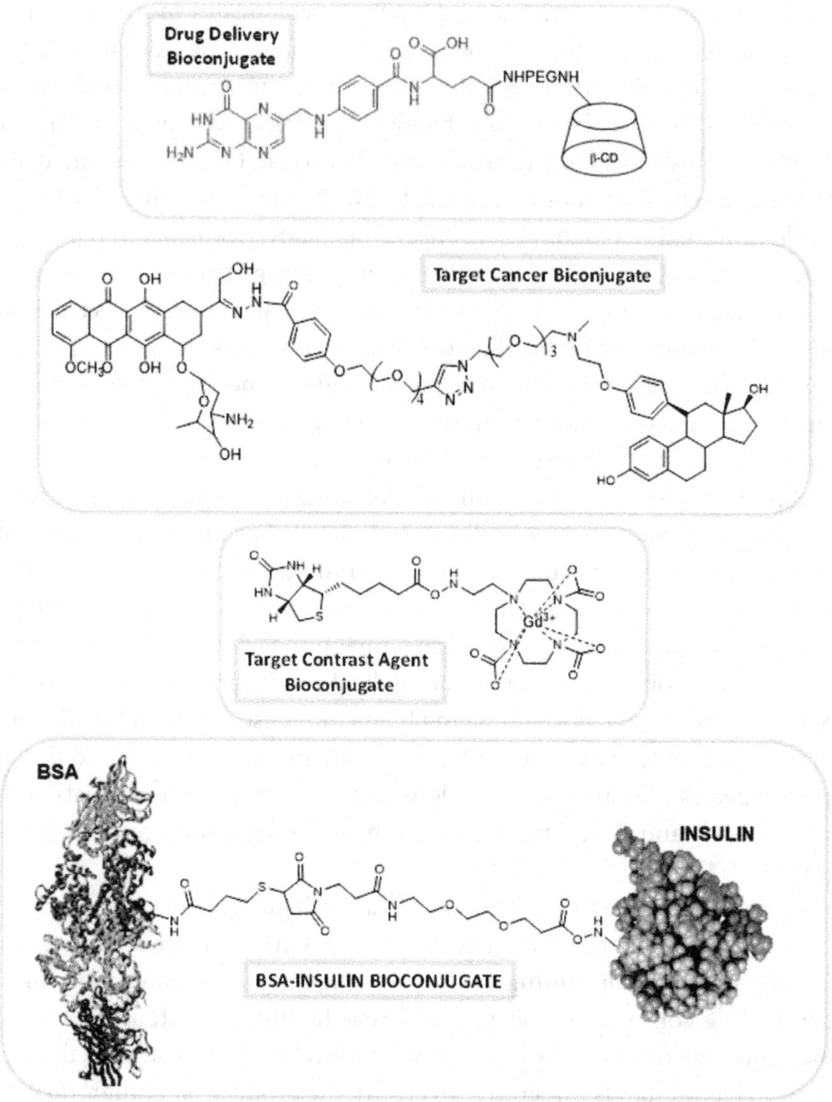

FIGURE 1.4 Selected examples of bioconjugates for medicinal applications.

Source: Adapted from [12,14,16,18].

biological and synthetic components, the immune system may perceive the synthetic molecule as foreign, triggering an immune response. This immune recognition can compromise the efficacy of the bioconjugate and amplify the potential for side effects. Addressing these challenges requires

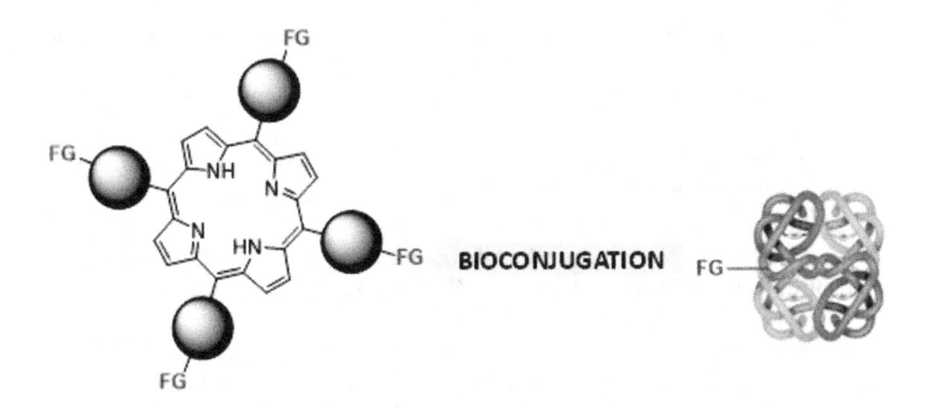

FIGURE 1.5 General structure of porphyrin–biomolecule bioconjugation.

extensive research and development efforts to enhance stability, prolong shelf life, and minimize immunogenic responses, ensuring the safe and effective application of bioconjugates in clinical practice [1,2]. Despite the challenges and potential shortcomings associated with bioconjugates, their undeniable benefits ensure that these molecules will continue to play a pivotal role in medicine and medicinal chemistry. The advantages they offer outweigh the limitations, making them a valuable asset in various applications.

The upmost relevance of porphyrins in both therapeutics and diagnosis, as well as their straightforward molecular modeling, has led us to select these molecules as relevant pedagogical models for teaching the fundamentals of bioconjugate synthesis. Thus, this book mainly concentrates on the synthesis of *meso*-aryl porphyrins and strategies to prepare its bioconjugates using biocompatible synthetic polymers (dendrimers, PEG) or biopolymers (carbohydrates, proteins). Porphyrins belong to a class of organic molecules characterized by an aromatic ring-like structure comprised of four pyrrole units linked by four methine bridges. These are aromatic molecules known by easy synthesis and functionalization, which are key points to promote its bioconjugation (**Figure 1.5**).

REFERENCES

[1] Sunasse, R., Narain, R., Covalent and Noncovalent Bioconjugation Strategies, in *Chemistry of Bioconjugates: Synthesis, Characterization, and Biomedical Applications* (Chapter 1), edited by Narain, R., John Wiley & Sons, Inc., **2014**, 3–75.

[2] Hermanson, G. T., *Bioconjugate Techniques*, Elsevier Inc., **2008**, 3–1202.

[3] Wuts, P. G. M., Greene, T. W., *Greene's Protective Groups in Organic Synthesis*, John Wiley & Sons, **2006**.

[4] Sartori, G., Ballini, R., Bigi, F., Bosica, G., Maggi, R., Righ, P., *Chem. Rev.*, **2004**, *104*, 199.

[5] Sureshbabu, V. V., Narendra, N., Protection Reactions, in *Amino Acids, Peptides and Proteins in Organic Chemistry*, edited by Hughe, A. B., Vol. 4, John Wiley & Sons, Inc., **2011**, 1.

[6] Isidro-Llobet, A., Alvarez, M., Albericio, F., *Chem. Rev.*, **2009**, *109*, 2455.

[7] Bird, R. E., Lemmel, S. A., Yu, X., Zhou, Q. A., *Bioconjug. Chem.*, **2021**, *32*, 2457.

[8] Ramstrom, O., The Royal Swedish Academy of Sciences, **2022**.

[9] Agrahari, A. K., Bose, P., Jaiswal, M. K., Rajkhowa, S., Singh, A. S., Hotha, S., Mishra, N., Tiwari, V. K., *Chem. Rev.*, **2021**, *121*, 7638.

[10] Kolb, H. C., Finn, M. G., Sharpless, K. B., *Angew. Chem., Int. Ed. Engl.*, **2001**, *40*, 2004.

[11] Eras, A., Castillo, D., Suárez, M., Vispo, N. S., Albericio, F., Rodriguez, H., *Front. Chem.*, **2022**, *10*, 889083.

[12] Caliceti, P., Salmaso, S., Semenzato, A., Carofiglio, T., Fornasier, R., Fermeglia, M., Ferrone, M., Pricl, S., *Bioconju. Chem.*, **2003**, *14*, 899.

[13] Wadhawan, A., Chatterjee, M., Singh, G., *Int. J. Mol. Sci.*, **2019**, *20*, 5243.

[14] Dao, K.-L., Sawant, R. R., Hendricks, J. A., Ronga, V., Torchilin, V. P., Hanson, R. N., *Bioconju. Chem.*, **2012**, *23*, 785.

[15] Yang, L., Fang, W., Ye, Y., Wang, Z., Hu, Q., Tang, B. Z., *Mater. Chem. Front.*, **2019**, *3*, 1335.

[16] Mishra, A., Pfeuffer, J., Mishra, R., Engelmann, J., Mishra, A. K., Ugurbil, K., Logothetis, N. K., *Bioconju. Chem.*, **2006**, *17*, 773.

[17] Yu, C., Xi, J., Li, M., A., M., Liu, H., *Bioconju. Chem.*, **2018**, *29*, 719.

[18] Rao, S. S., Somayaji, Y., Kulal, A., *ACS Omega*, **2022**, *7*, 5131.

CHAPTER 2

Synthesis of Porphyrins[1]

2.1 INTRODUCTION

Porphyrins and their reduced derivatives (chlorins and bacteriochlorins) constitute a naturally occurring class of tetrapyrrolic macrocycles that play a vital role in biological systems and in medicine. They consist of four pyrrole rings connected by four methine (=CH-) bridges, located at their 5, 10, 15, and 20 positions (**Figure 2.1**), which are referred as *meso* positions, according to *Fischer* nomenclature [1]. The 2, 3, 7, 8, 12, 13, 17, and 18 positions used in IUPAC nomenclature [2] are also called β-pyrrolic in *Fischer* designation. As highlighted in **Figure 2.1**, although they possess in total 26 π electrons, only 18 of them are involved in a delocalized π system [3]. This leaves two double bonds at positions 7–8 and 17–18 which do not participate in this delocalization and thus possess a chemical reactivity similar to that of alkenes. Indeed, one or both of these positions can be readily reduced, originating chlorins and bacteriochlorins, respectively.

Despite their apparently complex chemical structure, porphyrins and chlorins are ubiquitous in nature. This is surely a consequence of their unique properties, which have given a competitive advantage to many living organisms over the course of evolution. Nowadays, they play major biological roles as chromophores, molecular carriers, and enzymatic catalysts [4]. In **Figure 2.2**, two of the most paradigmatic examples of a natural porphyrin and chlorin are presented.

Heme is a prosthetic group present in many proteins, such as hemoglobin and myoglobin. Its ability to coordinate and release O_2 through its metal center allows oxygen distribution from lungs to peripheral tissues. Besides hemoglobin, heme groups are also present in cytochromes,

DOI: 10.1201/9781003119265-2

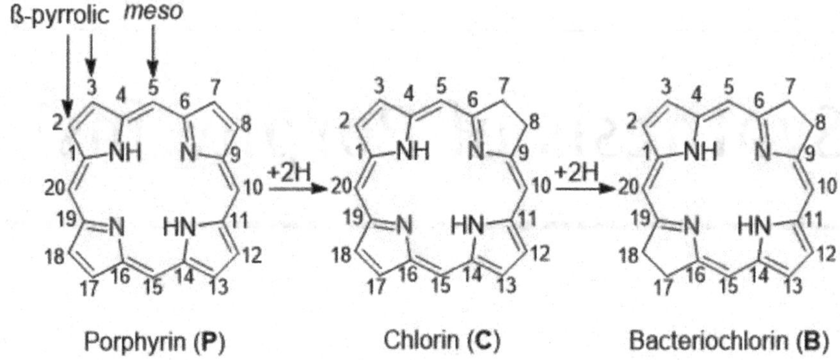

FIGURE 2.1 Generic structure of a porphyrin, chlorin, and bacteriochlorin containing the preferred IUPAC numbering, as well as the Fischer designations *meso* and β-pyrrolic positions. Highlighted in blue is the 18-π electron delocalization system.

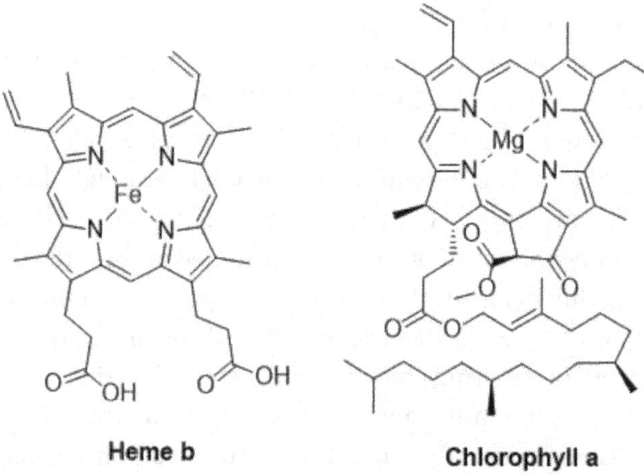

FIGURE 2.2 Examples of two of the most relevant tetrapyrrolic macrocycles found in nature.

catalases, and peroxidases, which are some of the most important redox enzymes that contribute to the homeostasis of organisms through the metabolization of endogenous molecules, as well as xenobiotics [5]. Their biological function is controlled by the Fe axial ligand of each heme-containing protein. On the other hand, chlorophylls are chlorins with intense green pigments responsible for photosynthesis in plants and some bacteria [6]. Through sunlight absorption, they can initiate an electron transfer chain that will culminate in the conversion of H_2O into O_2, and CO_2 into sugars. It is worth highlighting that naturally occurring porphyrins and derivatives are exclusively substituted at their β-pyrrolic positions. This is in clear contrast with the majority of synthetic porphyrins, which are substituted at their *meso* positions [7]. Indeed, the biosynthetic mechanisms associated with β-substituted porphyrins vastly differ from the ones developed by chemists over the last few decades, which will be discussed in more detail in this chapter.

The multitude of different roles that porphyrins perform in nature has motivated scientists to conduct an intensive search for finding new synthetic methods [8,9,10,11,12,13] and applications for tetrapyrrolic macrocycles. Such applications include energy conversion (solar cells) [14], catalysis [15,16,17], and chemical sensors [18,19]. Perhaps one of their most relevant applications is in medicinal chemistry, either as photosensitizers for photodynamic inactivation of cancer [20,21,22], microorganisms [23,24,25,26,27,28], and molecular imaging [29,30,31,32,33,34,35]. Some of the clinically relevant porphyrin derivatives (**Figure 2.3**) include temoporfin (chlorin approved for treatment of head and neck cancer) [36], padeliporfin (bacteriochlorin approved for treatment of pancreatic cancer) [37], and redaporfin (bacteriochlorin in clinical trials for treatment of head and neck cancer) [38]. As can be seen by their chemical structures, these porphyrin derivatives can either be purely synthetic *meso*-aryl-substituted (temoporfin and redaporfin) or hemi-synthetic, with substituents also in β-pyrrolic positions (padeliporfin).

The considerable chemical space associated with these tetrapyrrolic macrocycles offers an incredible advantage in this field, since their biological, physicochemical (e.g., water solubility, amphiphilicity), photophysical, and photochemical properties can be modulated. Some of the emerging challenges in their medical applications are to ensure enough biocompatibility and affinity to target cells. Therefore, their conjugation with biomolecules like peptides and sugars or synthetic polymeric materials like PEG is a very promising approach, as will be discussed with specific examples in the following chapters of this book. Thus, in this chapter we will describe the main synthetic methodologies for the synthesis and functionalization of the selected *meso*-substituted tetrapyrrolic macrocycles that will be used in the following sections to prepare bioconjugates.

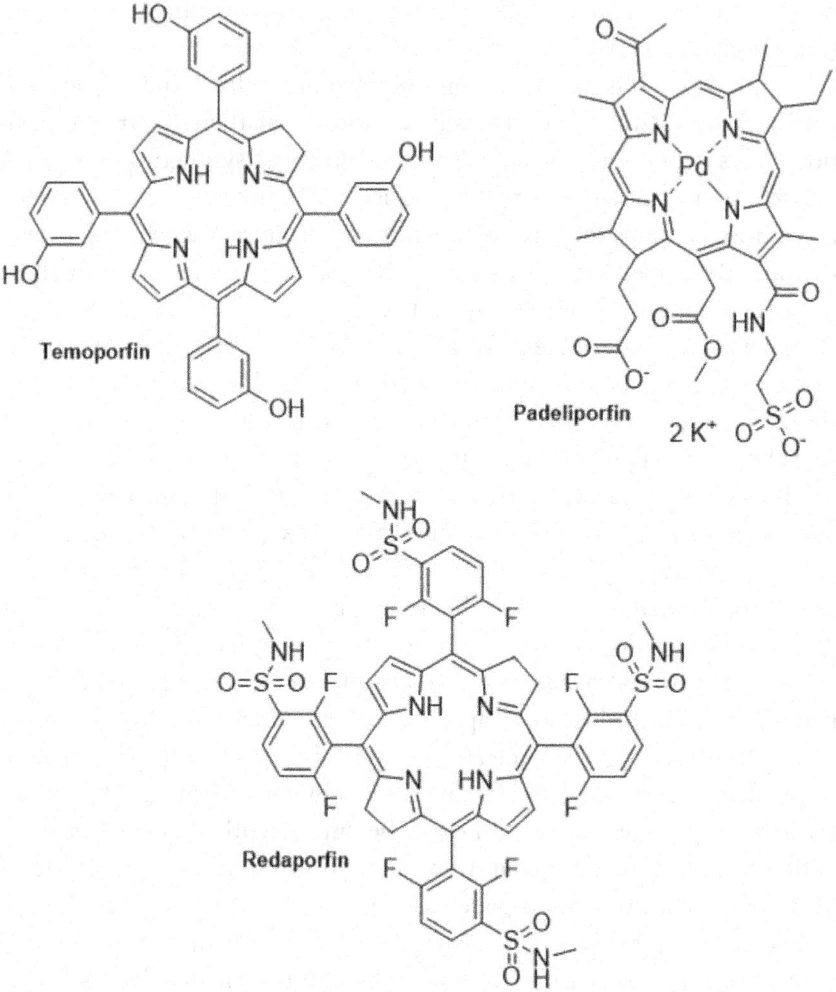

FIGURE 2.3 Examples of some landmark porphyrin derivatives, chlorins, and bacteriochlorins used in the photodynamic therapy of cancer (PDT).

2.2 ONE-POT SYNTHETIC APPROACHES FOR THE SYNTHESIS OF *MESO*-ARYL PORPHYRINS

Since the early 20th century, several methods have been reported for the one-pot synthesis of *meso*-aryl porphyrins. Typically, this reaction entails the condensation reaction of pyrrole with an aldehyde, in acidic conditions and in the presence of an oxidant. Before proceeding to the analysis of the reaction mechanism, it is important to understand why pyrrole reacts

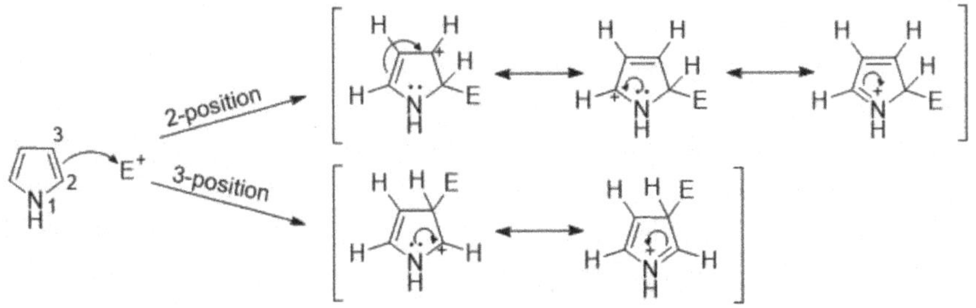

SCHEME 2.1 Resonance structures of reaction intermediates in the aromatic electrophilic substitution on 2 and 3 positions of pyrroles.

almost exclusively at its 2 position. In a typical aromatic electrophilic substitution reaction (**Scheme 2.1**), the nucleophilic attack of pyrrole to an electrophile (E^+) can theoretically occur from its 2 or 3 positions. This results in a positively charged non-aromatic intermediate state, which is stabilized by electronic delocalization inside the pyrrole ring. When the electrophile is added to the 2 position, this intermediate state is stabilized by three main resonance structures, while only two resonance structures stabilize the intermediate, resulting from an addition to the 3 position. In other words, the intermediate resulting from an attack at the 2 position can more easily accommodate the positive charge, resulting in a higher stability, and consequently lowering the activation energy barrier, which ultimately leads to a higher reactivity.

After these initial considerations on the reactivity of pyrrole, the reaction mechanism associated with the acid-catalyzed porphyrin synthesis is depicted in **Scheme 2.2**.

The synthetic route for symmetric *meso*-substituted porphyrins starts with the activation of the aldehyde through protonation of its oxygen atom or reaction with a Lewis acid. This facilitates the nucleophilic attack at the 2 position of pyrrole to the aldehyde carbon, with consequent formation of a carbinol intermediate (**A**). Then, the regeneration of pyrrole's aromaticity occurs through the loss of a proton from this 2 position (**B**). The protonation of the hydroxyl group (**C**) allows the nitrogen-assisted elimination of water, with consequent formation of a double bond (**D**). The conjugate addition of another pyrrole to the formed double bond allows the formation of dipyrromethane (**E**), which encompasses two pyrrole units linked by a methine bridge. Further reaction with another two aldehyde and two pyrrole molecules will eventually lead to the formation

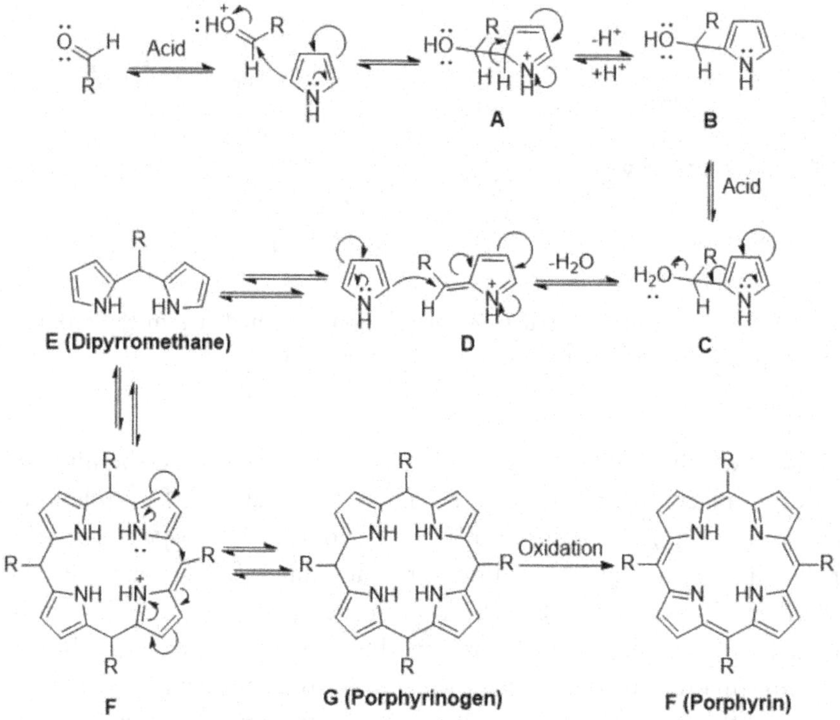

SCHEME 2.2 Reaction mechanism of the synthesis of symmetric *meso*-substituted porphyrins through condensation of an aldehyde with pyrrole.

of intermediate **F**, which comprises four pyrrole units. Alternatively, the condensation of two dipyrromethane units with two aldehyde molecules was also observed by NMR in the synthesis of *meso*-tetra-alkyl porphyrins [39]. At this point, an intramolecular ring closure can occur, leading to the formation of porphyrinogen (**G**). Alternatively, intermediate **F** can continue its polymerization process, leading to the formation of linear pyrrolic-based oligomers as undesired subproducts, which constitute the major drawback of *meso*-substituted porphyrin synthesis. It is worth mentioning that the formation of porphyrinogen is reversible, and thus, ring opening may occur, leading to more polymerization side products. However, the oxidation of porphyrinogen to porphyrin (**F**) is non-reversible, a consequence of the improved thermodynamic stability conferred by the established 18-π electron aromatic system. This oxidation can be done by atmospheric oxygen (*Adler-Longo*), nitrobenzene/air (*Gonsalves and Pereira*), or high-potential quinones (*Lindsey*).

Over the years, many important contributions towards the synthesis of *meso*-substituted porphyrins have been made aiming to reduce the formation of polymers and favoring closure and oxidation. In the following subsections, a more detailed overview on these landmark porphyrin synthetic methods will be given, which will include optimizations of the reaction solvent, reagent concentrations, acid strength, and oxidation methodologies.

2.2.1 Rothemund Method

Despite its limited use nowadays, in view of improved methodologies, the *Rothemund method* for the synthesis of *meso*-aryl porphyrins constitutes one of the most important milestones in the development of porphyrin chemistry. It was first reported by *Paul Rothemund* in 1935 [40] for the synthesis of a set of porphyrins, but with quite low isolated yields. This methodology was only applied for the synthesis of 5,10,15,20-tetraphenylporphyrin (TPP) in 1936 [41], and later, in 1941, newly optimized conditions were reported (**Scheme 2.3**) [42].

The synthesis of TPP was accomplished using equimolar quantities of pyrrole and benzaldehyde dissolved in pyridine (~3 M concentration) and placed in a sealed tube. The main problem was that the air was replaced by nitrogen and the tube was heated to 220°C in a furnace for 48 h. After a slow cooling process, this porphyrin crystallized directly from the reaction mixture, which was then filtered, giving TPP in 6–7% yield. The harsh

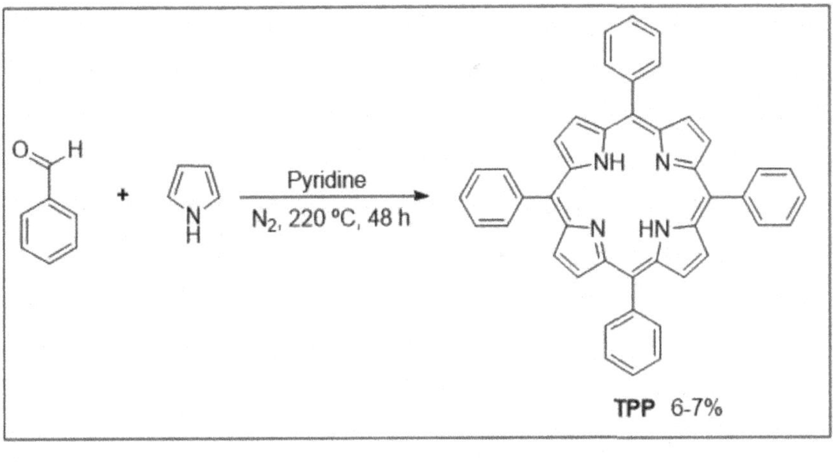

SCHEME 2.3 Synthesis of TPP using the Rothemund method [42].

reaction conditions employed by *Rothemund* and the absence of oxidant meant that only few substituted benzaldehydes could be successfully converted to the corresponding porphyrins [43]. This issue, combined with the low yields and contamination with the corresponding chlorin, induced the search for improved methods that could rely on efficient catalysts, rather than a "brute-force" approach. Despite the poor yield and difficulty in reaction reproducibility, the initial *Rothemund* one-pot synthesis of porphyrins is still referenced nowadays and has more than 1,000 combined citations from its landmark papers, according to Google Scholar.

2.2.2 Adler–Longo Method

The methodology first reported by *Adler and Longo* in 1964 was a game changer, as it introduced significant improvements to the pioneering *Rothemund* synthetic methodology. Pyridine was substituted by propionic or acetic acid as solvent, and the reaction was carried in an open vessel, exposed to atmospheric oxygen. By using an acid as both solvent and catalyst, the aldehydes are more efficiently activated, allowing the reaction to proceed under lower temperatures and for shorter times. Additionally, the open-vessel approach allows a continuous supply of atmospheric oxygen, which aids in the oxidation of the porphyrinogen to porphyrin, thus avoiding ring opening with formation of polymerization side products. In the specific case of TPP synthesis (**Scheme 2.4**), equimolar quantities of pyrrole and benzaldehyde (0.27 M concentration) are added to refluxing

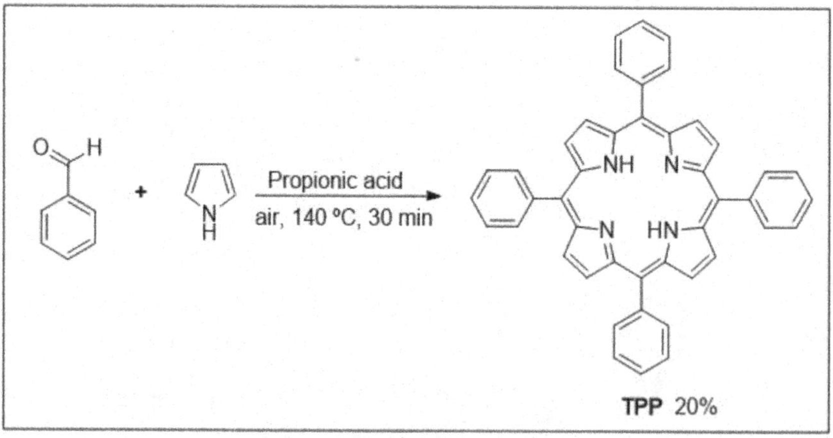

SCHEME 2.4 Synthesis of TPP using the Adler–Longo method [44].

propionic acid (~140°C). After 30 min, the reaction is cooled and the product is filtered, giving TPP in 20% yield, always with about 3–10% contamination with the corresponding chlorin [44].

In **Table 2.1**, some examples of porphyrins intended for bioconjugation, synthesized by the *Adler–Longo* method, are presented. For instance, the synthesis of bioconjugates between dendrimers and porphyrins (**Chapter 3**) requires the preparation of porphyrins embedded with multiple hydroxyl groups to allow a branched dendrimer formation around the macrocycle. Typically, *meso*-hydroxyphenylporphyrins are

TABLE 2.1 Examples of Porphyrins Synthesized by the Adler–Longo Method, Aiming Bioconjugation

#	Structure	Yield	Ref
1	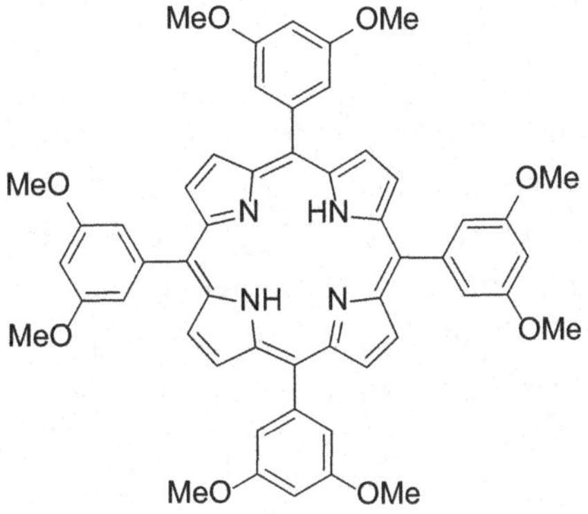 **P1**	53%	[47]
2	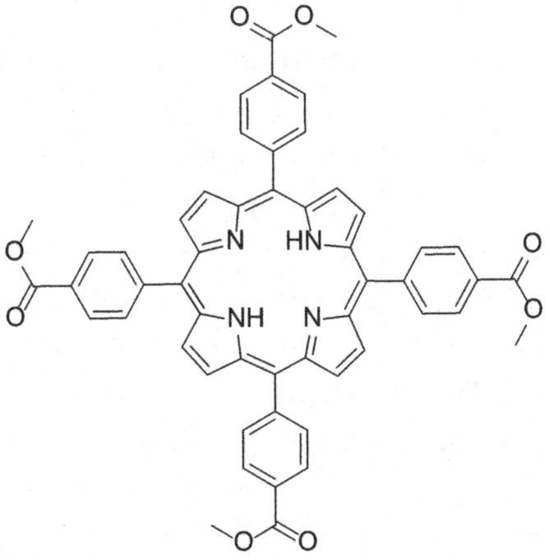 **P3**	55%	[48]

TABLE 2.1 (*Continued*) Examples of Porphyrins Synthesized by the Adler–Longo Method, Aiming Bioconjugation

#	Structure	Yield	Ref
3	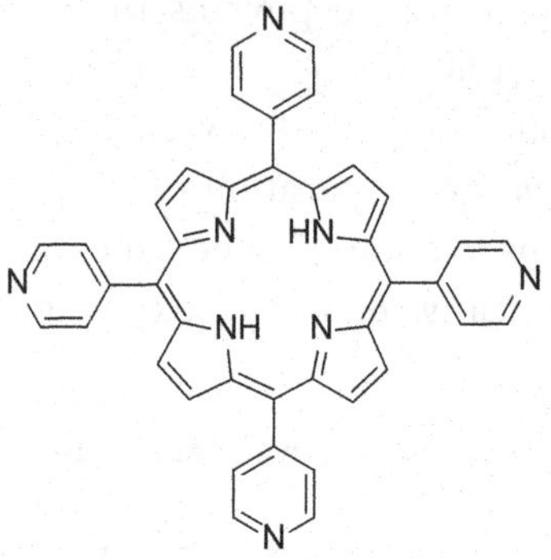 N; N; HN; N; NH; N; N; N **P6**	14%	[49]

not obtained directly from the corresponding hydroxybenzaldehyde due to low yields and cumbersome purification steps [45], and thus, they are prepared from protected methoxybenzaldehyde. As a selected example, *Moore* [46] synthesized a porphyrin–dendrimer conjugated that started with the synthesis of 5,10,15,20-tetrakis(3,5-dimethoxyphenyl)porphyrin (**P1**; **Table 2.1**, entry 1). This porphyrin can be obtained under the *Adler* methodology, as described earlier, in 53% yield. Then, O-demethylation, using BBr_3 as Lewis acid, in dichloromethane, affords the corresponding 5,10,15,20-tetrakis(3,5-dihydroxyphenyl)porphyrin in 94% yield (**Scheme 2.5**) [47].

The mechanism for this reaction (**Scheme 2.5**) entails the formation of a Lewis adduct between the oxygen and BBr_3, followed by elimination of Br^-. Then, demethylation is carried out through a nucleophilic attack of the Br- to the methyl group. Finally, addition of water breaks the O-B bond, with consequent formation of the free hydroxyl group, as well as HBr and $B(OH)_3$.

The preparation of a Pt(II) *meso*-carboxyphenyl-substituted porphyrin for dendrimer synthesis was described by *Odai* [51]. Following a similar strategy to that described for **P1**, these types of porphyrins containing free carboxyl groups are best prepared starting from an aldehyde containing protected carboxyl groups, such as methyl esters. For this specific compound, its synthetic route starts with 5,10,15,20-tetrakis(4-methoxycarbonylphenyl)porphyrin **P3**, previously reported through the reaction of pyrrole with methyl 4-formylbenzoate, under *Adler–Longo* conditions,

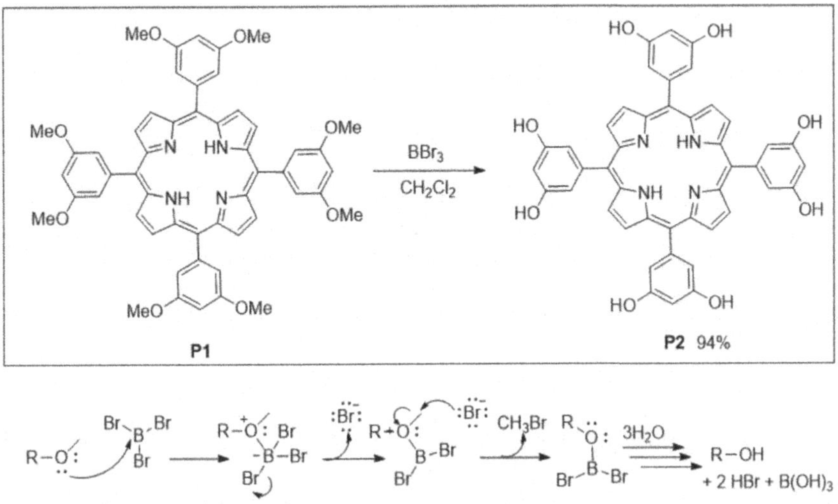

SCHEME 2.5 Synthesis of **P2** containing free hydroxyl groups for further preparation of dendrimers catalyzed by the Lewis acid BBr_3.

affording 55% yield (**Table 2.1**, entry 2) [48]. The presence of methyl ester groups greatly facilitates this porphyrin's work-up procedure, since it is not water-soluble and can also be purified by silica gel chromatography. The next step is the introduction of Pt(II) in the porphyrin's core, by using $PtCl_2$, in benzonitrile, at reflux temperate for 7 h. Again, the product can be easily purified by standard laboratory techniques such as silica gel chromatography, affording the desired Pt(II) complex **P4** with 54% yield. Finally, the hydrolysis of the methyl ester groups was carried out by dissolving KOH pellets in a porphyrin **P4** THF solution (containing 1% ethanol), and the reaction was carried out at room temperature for 3 h. Then, water was added and the reaction was stopped after 1 h by evaporation of the organic solvent. The crude was acidified by adding concentrated HCl, causing the protonation of the carboxylic groups, and then washed with water, affording the Pt(II) 5,10,15,20-tetrakis(4-carboxyphenyl)porphyrin **P5** in 90% yield (**Scheme 2.6**) [52]. The linking of this porphyrin to dendrimers for use as cellular oxygen sensors will be discussed in **Chapter 3**.

The *Adler* methodology can also be used to prepare 5,10,15,20-tetrakis(4-pyridyl)porphyrin **P6**, in 14% yields (entry 3). Then, *Kawakami* reacted **P6** with 3-bromopropylamine hydrobromide in reflux temperature for 72 h, obtaining **P7**, bearing four free amino groups (**Scheme 2.7**). This tetracationic porphyrin has an improved water solubility and biocompatibility,

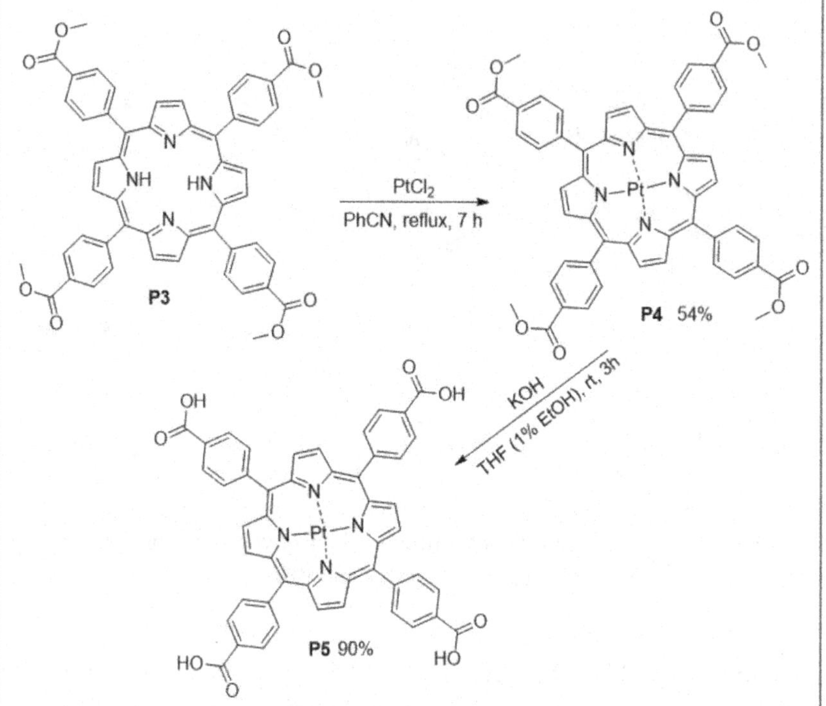

SCHEME 2.6 Synthesis of the Pt(II) porphyrin **P5** containing free carboxyl groups for further preparation of dendrimers.

while the amino groups can be used for bioconjugation with aldehydes from sugars, as will be discussed in **Chapter 5** [53].

Despite the important contributions towards the synthesis of *meso*-aryl porphyrins, the *Adler–Longo* method is inefficient in the synthesis of porphyrins containing halogenated bulky groups on the *ortho*-positions of the phenyl rings. This, combined with the frequent chlorin contaminations, has fueled the development of alternative one-pot methodologies, such as the *Gonsalves and Pereira* nitrobenzene method, described in the next section [54].

2.2.3 Nitrobenzene Method

The nitrobenzene method was developed by *Gonsalves and Pereira* and reported in 1991 [54], replacing the solvent used in the *Adler* method (an organic acid) by a mixture of acetic acid/nitrobenzene (2:1). The reaction is also carried out under aerobic conditions, at temperatures up to 150°C,

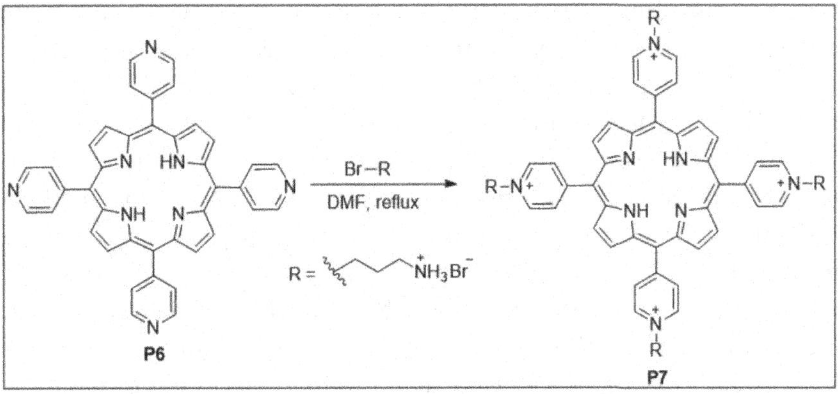

SCHEME 2.7 Synthesis of the cationic tetra-pyridyl **P7** containing free amino groups for further bioconjugation to sugars.

with the high reagent concentrations (0.2 M) and reaction times up to 1 h, resulting in a multitude of different *meso*-aryl porphyrins obtained. Nitrobenzene plays a key role both as solvent and as oxidant of the reaction intermediates, porphyrinogen and/or chlorin, to the corresponding porphyrin. This method has led to significant improvements on the final porphyrin's overall yields (**Scheme 2.8**). While TPP yield is the same to that of the *Adler* method (20%), hindered porphyrins containing groups in the *ortho*-positions, such as 2-nitrophenyl (20%), 2-methoxyphenyl (15%), 2-chlorophenyl (8.5%), and 2,6-dichlorophenyl (5%) groups, were obtained in higher yields than in previously reported methodologies [55,56,57,58]. Additionally, the purity and work-up procedures of the porphyrin products were improved, since this type of porphyrins is, in general, directly

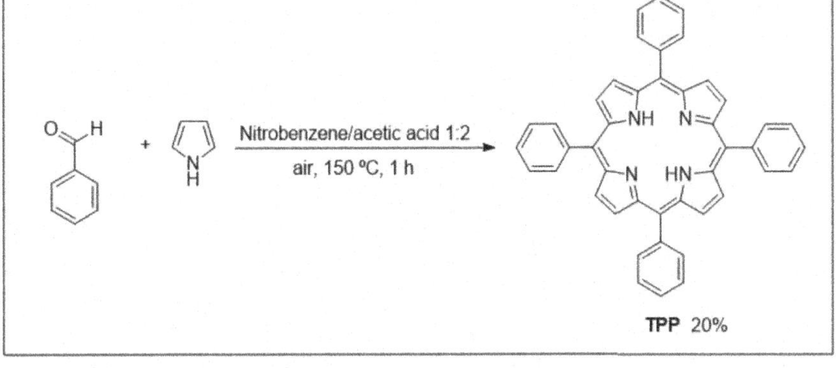

SCHEME 2.8 Synthesis of *meso*-aryl porphyrins using the nitrobenzene method.

isolated from direct precipitation from the reaction medium, without almost any contamination with the corresponding chlorins [54].

Later, *Pereira* [59] showed that the addition of an aluminosilicate such as NaY, a Lewis acid, to the reaction mixture could improve the isolated yields of some porphyrins. Furthermore, this solid catalyst can be filtered off and reused, which highlights the sustainability of this method. This nitrobenzene one-pot synthesis of *meso*-aryl porphyrins allowed for the first time the synthesis of porphyrins with bulky substituents on the *ortho*-positions of the phenyl ring, but still with low yields [10]. More recently, the same authors described the use of water, under microwave irradiation, as a promising sustainable alternative for the preparation of *meso*-aryl porphyrins [60].

In **Table 2.2**, some examples of porphyrins intended for bioconjugation, synthesized by the nitrobenzene method, are presented. The synthesis of **P3**, previously mentioned in **Table 2.1** (entry 2, **Table 2.2**), was also achieved under the nitrobenzene method, albeit with lower yield (20%), when compared with the *Adler–Longo* method (55%).

On the field of PEG–porphyrin bioconjugates, *Pereira* [34] reported the synthesis of 5,10,15,20-tetrakis(pentafluorophenyl)porphyrin (**P9**) and its conjugation with PEG500. This base porphyrin was obtained under Nitrobenene and Nitrobenzene/NaY methodologies, in 8 and 17% yields, respectively (entry 2, **Table 2.2**). Overall, the use of the NaY acidic aluminosilicate in conjunction with the nitrobenzene method constitutes

TABLE 2.2 Examples of Porphyrins Synthesized by the Nitrobenzene Method, Aiming Bioconjugation

#	Structure	Method	Yield	Ref
1	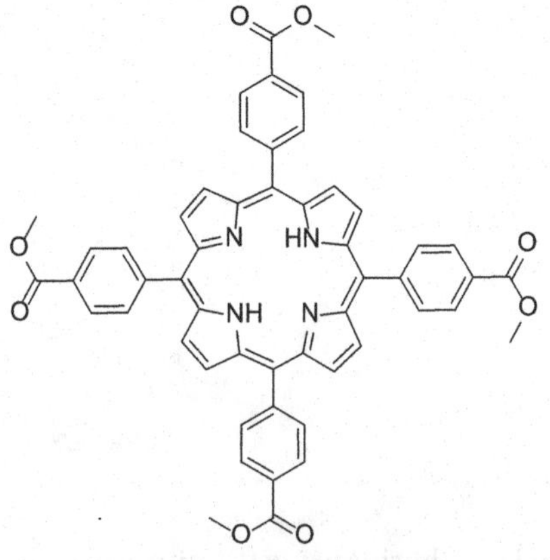 O, O, N, HN, NH, N, O, O, O, O, O, O **P3**	Nitrobenzene	20%	[61]

TABLE 2.2 (*Continued*) Examples of Porphyrins Synthesized by the Nitrobenzene Method, Aiming Bioconjugation

#	Structure	Method	Yield	Ref
2	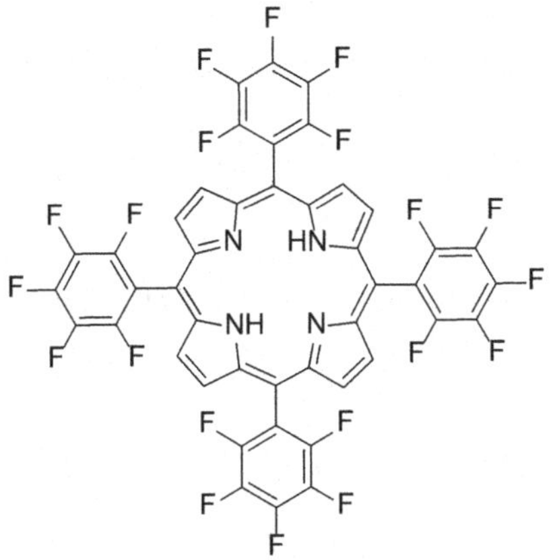 **P9**	Nitrobenzene Nitrobenzene + NaY	8% 17%	[62] [59]
3	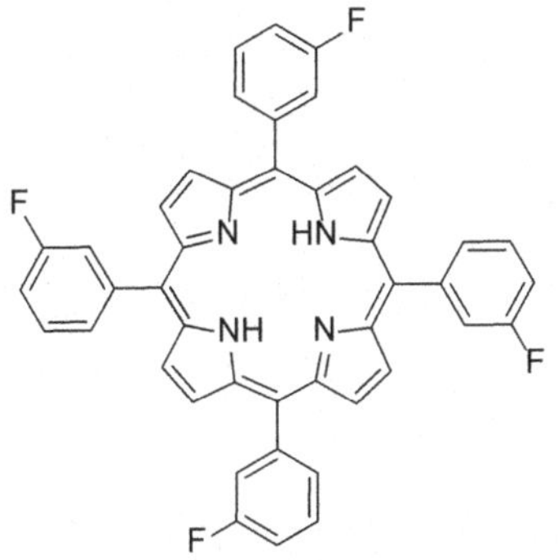 **P10**	Nitrobenzene	30%	[63]
4	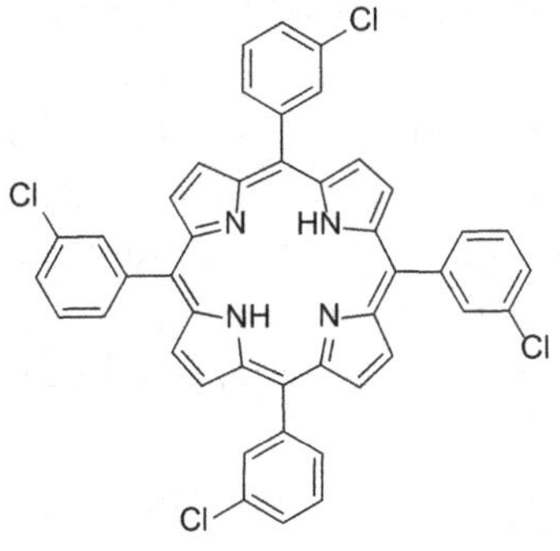 **P11**	Nitrobenzene	22%	[64]

a twofold improvement in yield over the classic nitrobenzene method. **P9** is a remarkably versatile intermediate for bioconjugation, since its pentafluorophenyl ring is highly activated towards aromatic nucleophilic

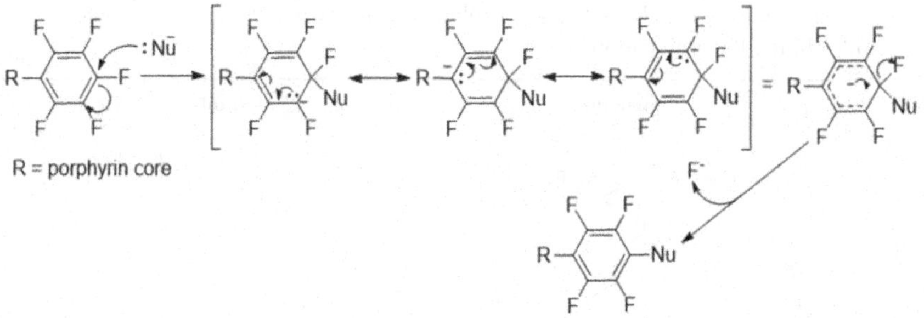

SCHEME 2.9 Mechanism for the nucleophilic aromatic substitution (S_NAr) in pentafluorophenyl rings by a nucleophile (Nu).

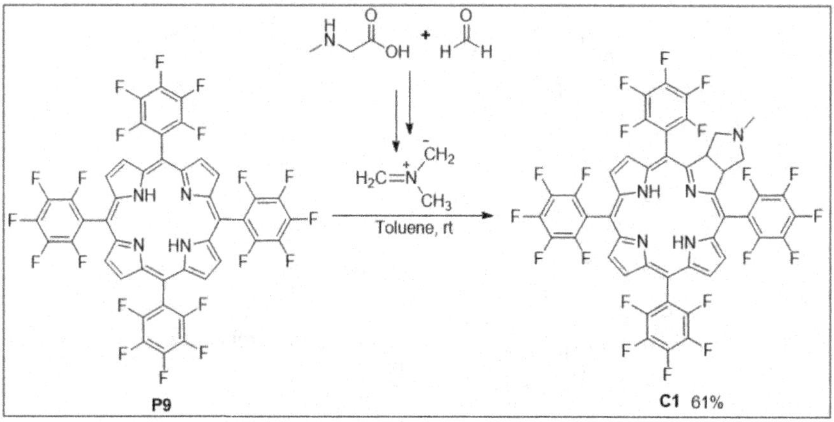

SCHEME 2.10 Synthesis of **C1** through a 1,3-dipolar cycloaddition.

substitutions (S_NAr). Given that fluorine atoms are *meta*-orienting towards S_NAr reactions, when considering all individual contributions in the pentafluorophenyl ring, it is possible to conclude that the carbon at the *para* position relative to the macrocycle is the most reactive. Indeed, experimental data from multiple authors using different nucleophiles (alcohols [65], amines [66], sulfonamides [67], thiols [68], or organolithiums [69]) point out that the reaction occurs exclusively at this position (**Scheme 2.9**).

Silva described a modification of **P9** whereby a 1,3-dipolar cycloaddition with an azomethine ylide occurs in toluene, at room temperature, leading to the formation of the corresponding chlorin, **C1**, in 61% yield (**Scheme 2.10**) [70]. The reduction of porphyrin to chlorin allows a bathochromic shift of its absorption spectrum into the red/infrared region (phototherapeutic window), which is ideal for the preparation of bioconjugates aimed towards light-mediated medicinal applications.

Through the nitrobenzene synthetic methodology, 5,10,15,20-tetrakis(3-fluorophenyl)porphyrin (**P10**) and 5,10,15,20-tetrakis(3-chlorophenyl) porphyrin (**P11**) can also be obtained in 30% and 22% yields, respectively (entries 3 and 4, **Table 2.2**) [63,64]. These porphyrins can then be nitrated at the phenyl's 4 position for further use in bioconjugate chemistry, as will be discussed in Section 2.5.1.

2.3 TWO-STEP SYNTHETIC APPROACHES FOR SYNTHESIS OF *MESO*-ARYL PORPHYRINS

Despite the improvements introduced by *Adler* and nitrobenzene methodologies, the synthesis of *ortho*-substituted *meso*-aryl porphyrins was still a challenging synthetic issue. The pioneering work developed by *Gonsalvens and Pereira* in 1985 for the synthesis of *meso*-alkyl substituted porphyrins [39] was later extended by *Lindsey* in 1986 for the synthesis of *meso*-aryl porphyrins [71,72]. In both cases, the synthesis of porphyrins was carried out in two steps: (a) condensation of pyrrole with aldehyde under inert atmosphere, with consequent formation of the porphyrinogen; (b) oxidation of the porphyrinogen to porphyrin (**Scheme 2.11**). This two-step procedure employed chlorinated solvents (CCl_4, $CHCl_3$, or CH_2Cl_2) and strong acids (trifluoroacetic acid or a Lewis acid, such as BF_3). Typically, the condensation reaction (first step) was conducted at much milder conditions (usually room temperature) than the previously mentioned one-pot procedures but required considerably high dilutions of the pyrrole and aldehyde (0.01 M). These conditions provided a good balance between pyrrole polymerization and cyclization to give the porphyrinogen. Then, the second step entails the addition of an organic soluble oxidant such as *p*-chloranil or 2,3-dichloro-5,6-dicyano-1,4-benzoquinone (DDQ), or air/light, which oxidize the porphyrinogen to porphyrin, with negligible traces of chlorin contamination. The typical work-up procedure involves washing of the organic phase with alkaline aqueous solutions to remove traces of the employed quinones and its side products, followed by purification through silica or alumina column chromatography. In the case of TPP synthesis, the *Lindsey* methodology can give yields up to 50% [73], while for relevant *ortho*-phenyl-substituted porphyrins, yields in the range of 10–28% were obtained [74].

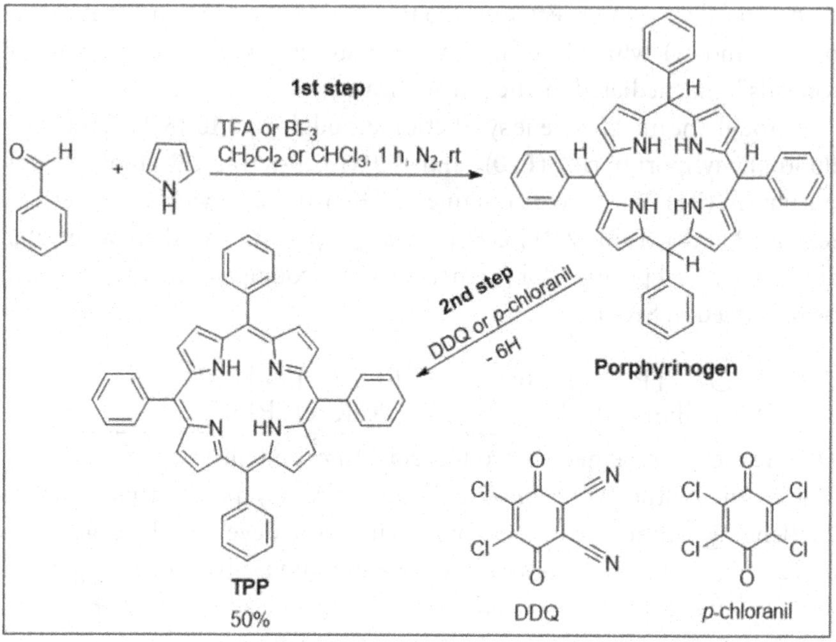

SCHEME 2.11 Synthesis of *meso*-aryl porphyrins using *Pereira* and *Lindsey's* two-step method [73].

In sum, there is no ideal method for the synthesis of *meso*-aryl porphyrins, and the choice of synthetic methodology is strongly dependent on the nature of the aldehyde. While the *Lindsey* method can achieve high yields for some porphyrins, the copious amounts of chlorinated solvents needed, combined with the undesired residues produced during porphyrinogen oxidation by quinones, are still concerning setbacks for its industrial-scale transposition.

In **Table 2.3**, some examples of porphyrins intended for bioconjugation, synthesized by the *Lindsey* method, are presented. For example, **P1**, **P3, P9,** and **P11**, described in the previous sections, were obtained in 52%, 40%, 41%, and 36% yields, respectively.

While the methods described so far in this chapter concern the synthesis of symmetric *meso*-tetra-aryl porphyrins, the synthesis of non-symmetric *meso*-aryl porphyrins with appropriate functional groups is a key issue for the preparation of tetrapyrrolic-based bioconjugates and will be discussed in the next sections.

TABLE 2.3 Examples of Porphyrins Synthesized by the *Lindsey* Method, Aiming Bioconjugation

#	Structure	Yield	Ref
1	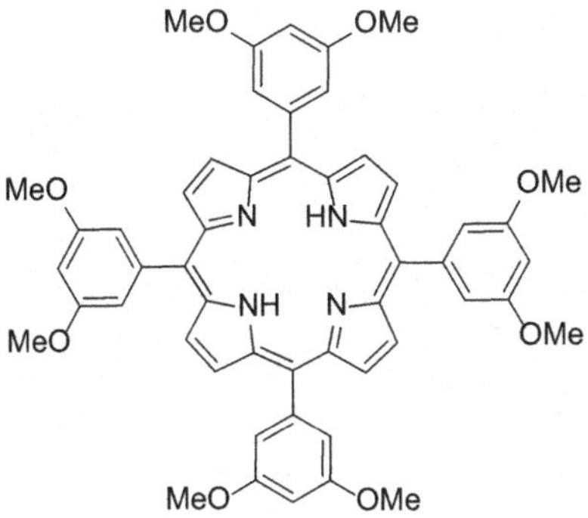 **P1**	52%	[75]
2	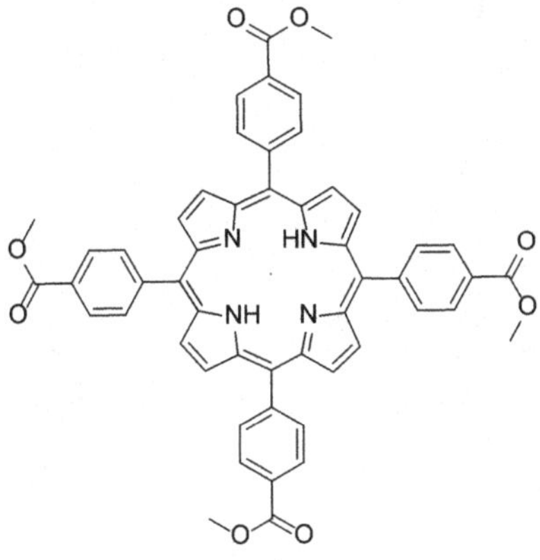 **P3**	40%	[76]
3	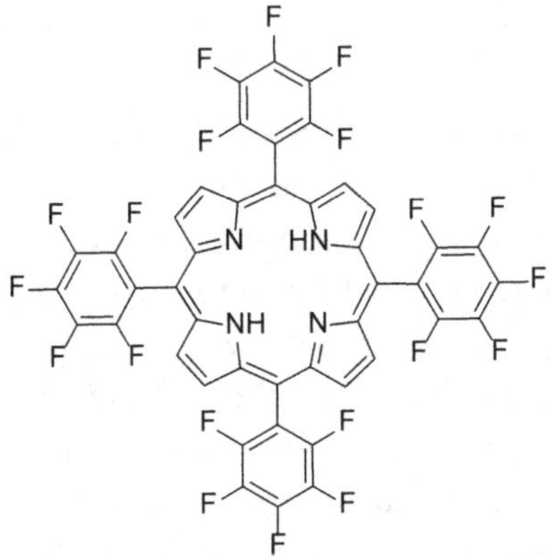 **P9**	41%	[77]

(*Continued*)

TABLE 2.3 (*Continued*) Examples of Porphyrins Synthesized by the *Lindsey* Method, Aiming Bioconjugation

#	Structure	Yield	Ref
4	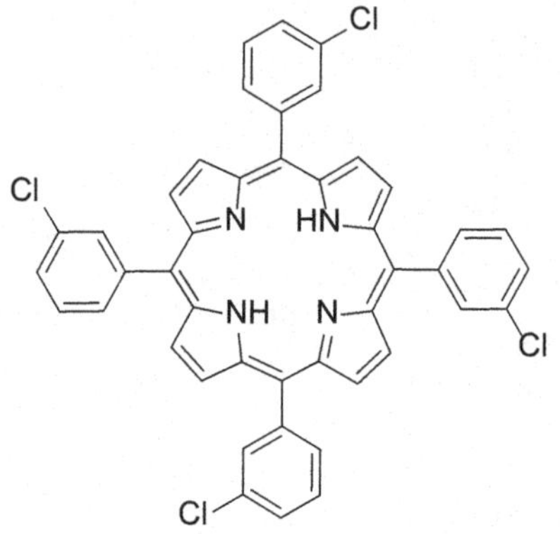 Cl, Cl, NH, N, N, HN, Cl, Cl **P11**	36%	[78]
5	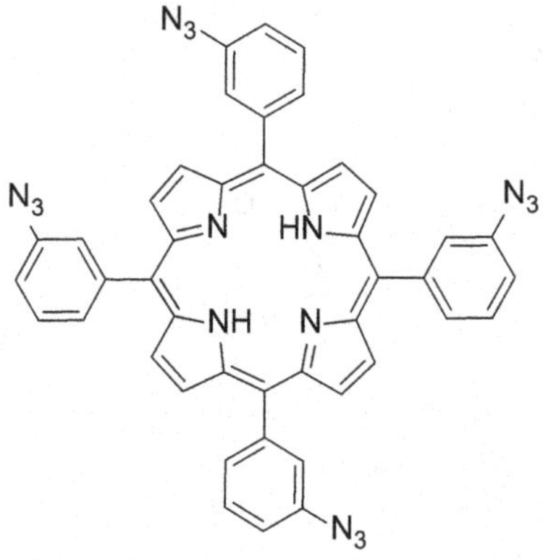 N_3, N_3, N, HN, NH, N, N_3, N_3 **P8**	31%	[79]

Umezawa described the synthesis of **P8** containing four azide groups under *Adler* conditions (entry 4), aimed for the preparation of bioconjugates with alkyne-containing peptides, through click chemistry. For this synthesis, the 3-azidobenzaldehyde was first prepared in a three-step methodology, starting from 3-aminobenzoic acid: (a) reduction of 3-aminobenzoic acid to 3-aminobenzyl alcohol with $LiAlH_4$; (b) synthesis of the corresponding diazonium salt (more details in Section 2.5.2), followed by substitution with the azide group, forming 3-azidobenzyl alcohol; (c) selective reduction of the benzyl alcohol to the corresponding aldehyde, using pyridinium dichromate, resulting in the desired

3-azidobenzaldehyde. The corresponding porphyrin was then obtained by reacting this aldehyde with pyrrole in dichloromethane, using BF_3 as acid catalyst and DDQ as oxidant, affording P8 in 31% yield [79].

2.4 SYNTHESIS OF NON-SYMMETRICAL PORPHYRINS

2.4.1 Mixed Aldehyde Condensation

One of the simplest methods to prepare non-symmetrical *meso*-aryl porphyrins is to carry out pyrrole condensation with a mixture of aldehydes (**Scheme 12**) [13,79].

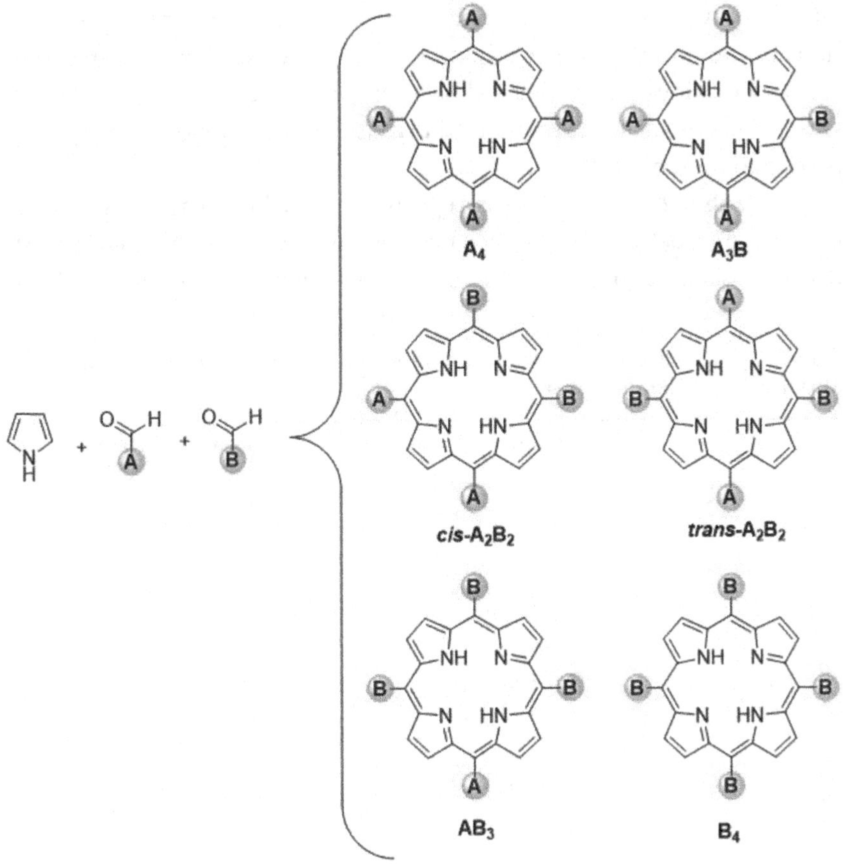

SCHEME 2.12 Theoretical *meso*-aryl porphyrins obtained through a mixed aldehyde (**A** and **B**) condensation.

For instance, when two different aldehydes are mixed (with substituents A and B), a total of six porphyrins can be theoretically obtained: A_4, A_3B, *cis*-A_2B_2, *trans*-A_2B_2, AB_3, B_4. Assuming that both aldehydes possess similar reactivities, the formation of A_2B_2-type porphyrins can be favored by using a 1:1 aldehyde ratio, which would yield approximately 6.25% of A_4, 25% of A_3B, 25% of *cis*-A_2B_2, 12.5% of *trans*-A_2B_2, 25% of AB_3, and 6.25% of B_4. If the goal is to favor the synthesis of a A_3B porphyrin, a 3:1 aldehyde ratio would give A_4 in 31% and A_3B in 42%. It is worth emphasizing that different aldehyde reactivities will change the expected outcome of these mixed aldehyde condensations, and thus, further optimizations on a case-by-case basis may be required to find the best aldehyde ratio. Despite the apparent simplicity of mixed aldehyde condensations, they often require laborious and time-consuming chromatographic separations of the products, especially if both aldehydes yield porphyrins with similar polarities [80]. Nevertheless, there are some examples of porphyrins for bioconjugation synthesized through this methodology, which are presented in **Table 2.4**.

P12 (**Table 2.4**, entry 1) was reported by *Lai* [81], where a statistical mixture of aldehydes (4-carboxybenzaldehyde and 4-acetamidobenzaldehyde) was used, based on the *Adler–Longo* methodology (no yield was

TABLE 2.4 Examples of Non-Symmetric Porphyrins Synthesized through Aldehyde Mixture, Aiming Bioconjugation

#	Structure	Method	Yield	Ref
1	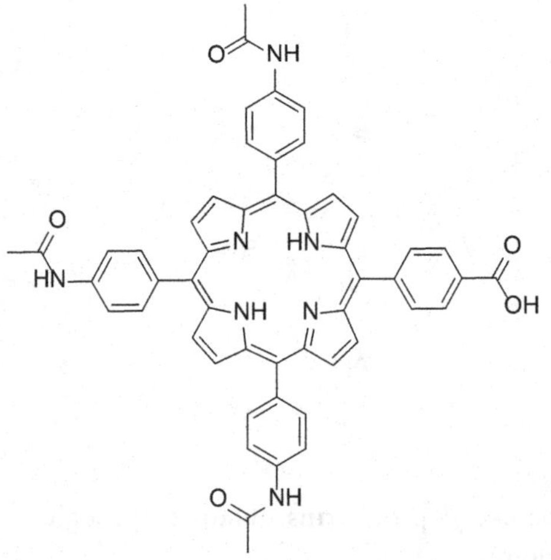 **P12**	*Adler-Longo*	Not reported	[81]

TABLE 2.4 (*Continued*) Examples of Non-Symmetric Porphyrins Synthesized through Aldehyde Mixture, Aiming Bioconjugation

#	Structure	Method	Yield	Ref
2	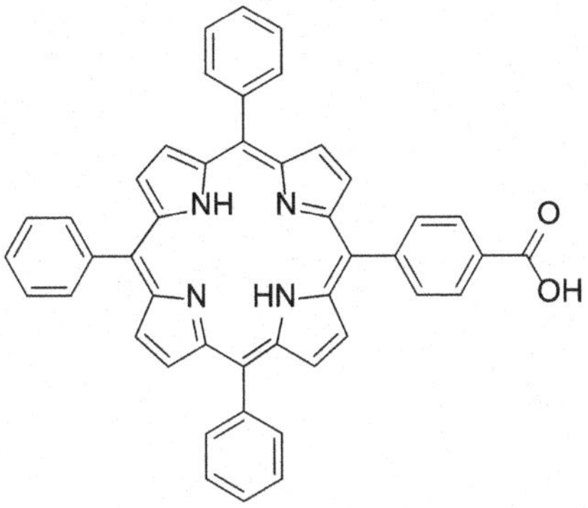 NH, N, N, HN, O, OH **P13**	*Lindsey*	12%	[82]
3	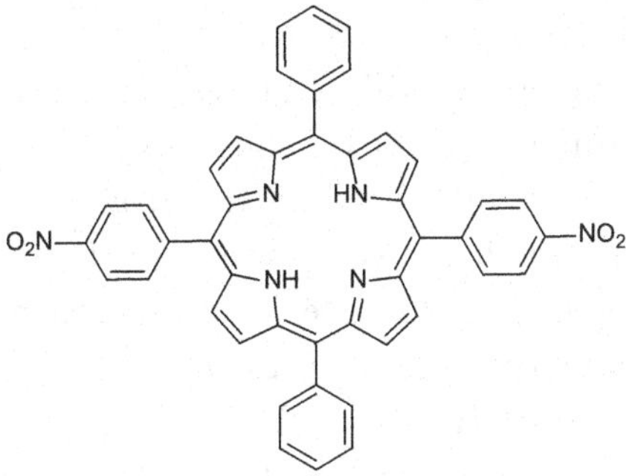 O_2N, N, HN, NH, N, NO_2 **P14**	*Lindsey*	12%	[83]
4	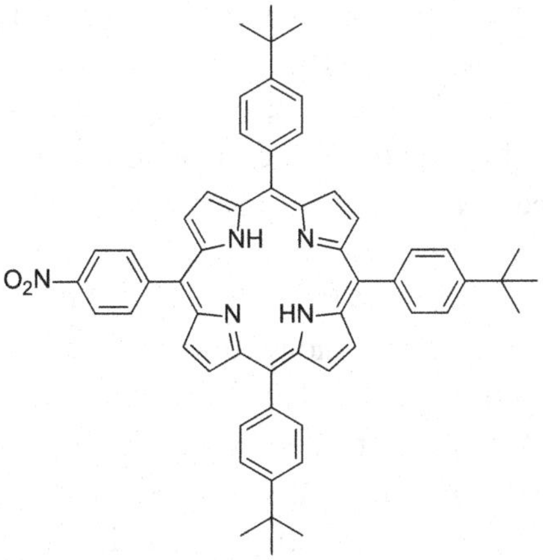 O_2N, NH, N, N, HN **P15**	*Lindsey*	11%	[84]

(*Continued*)

TABLE 2.4 (*Continued*) Examples of Non-Symmetric Porphyrins Synthesized through Aldehyde Mixture, Aiming Bioconjugation

#	Structure	Method	Yield	Ref
5	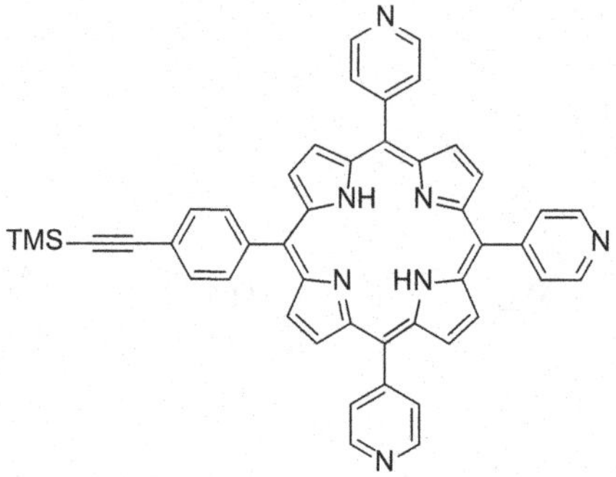 TMS; N; NH; N; N; HN; N; N; **P16**	*Lindsey modified*	8%	[85]

reported). The carboxylic acid functionality was used for linking to polyamidoamine (PAMAM) dendrimers, as will be presented in **Chapter 3**. Another example of a non-symmetric carboxyl-containing porphyrin is **P13 (Table 2.4**, entry 2), which was synthesized starting from pyrrole and a mixture of benzaldehyde with methyl 4-formylbenzoate. The resulting A_3B-type porphyrin containing one methyl ester group was then hydrolyzed to its carboxylic acid counterpart, in the presence of NaOH, yielding **P13** in 12% yield, after chromatographic purification in silica gel. This carboxyl-containing porphyrin was conjugated with glucosamine, as will be discussed in **Chapter 5**.

P14 and **P15** (**Table 2.4**, entries 3 and 4), containing nitro-aryl groups, were obtained in 12% and 11% yields, respectively, by mixed aldehyde condensation. These nitro groups can then be reduced to the corresponding amines, which can then be used in bioconjugation reactions. This specific synthetic step will be discussed in more detail in Section 2.5.

Feese reported the synthesis of the non-symmetric **P16** (entry 5, **Table 2.4**), containing one alkyne, protected with trimethylsilyl (TMS) and three pyridyl groups, under a two-step methodology, using xylene as solvent and salicylic acid as catalyst, under argon atmosphere. The resulting porphyrinogen was then oxidized to the corresponding porphyrin by atmospheric oxygen, yielding 8% of **P16**. The TMS group was then removed under basic conditions (K_2CO_3), in the presence of methanol, yielding **P17 (Scheme 2.13)** [85]. Mechanistically, it consists in a S_N2-type nucleophilic attack of the generated methoxy species to the Si atom, followed

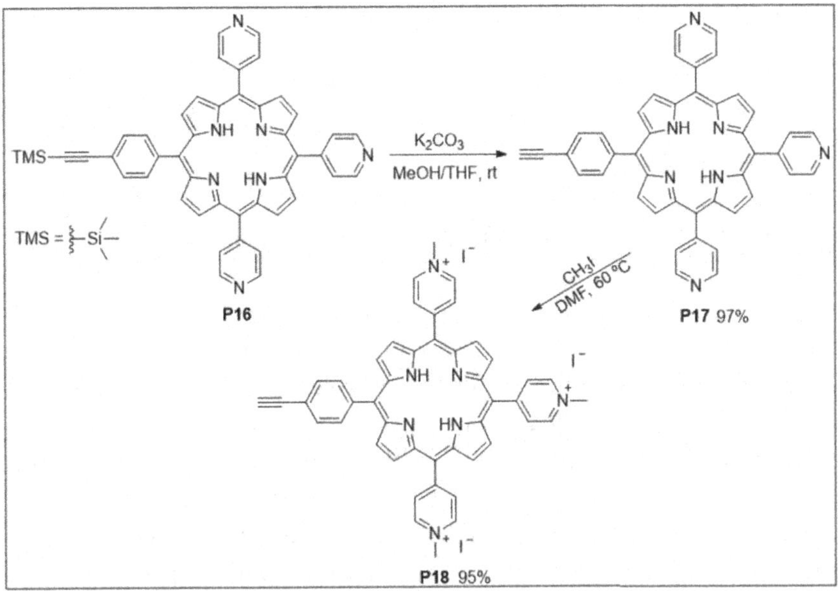

SCHEME 2.13 Synthesis of tri-cationic **P18** by TMS deprotection of **P16**, followed by cationization with iodomethane.

by cleavage of the Si-C_{sp} bond. Finally, the three pyridyl groups of **P17** were then alkylated with iodomethane, in DMF, forming the corresponding tri-cationic **P18**, which was then conjugated with a peptide for use in antimicrobial therapy, as will be presented in **Chapter 6**.

2.4.2 Dipyrromethane-Based Condensations

Over the last few decades, many contributions have been made on the development of strategies for a more rational and directed synthesis of non-symmetric *meso*-aryl porphyrins [7,86,87,88,89]. One of the key compounds for such reactions are dipyrromethanes **I**, which are intermediates in porphyrin synthesis (see mechanism in **Scheme 2.2**). Their syntheses [90,91] (**Scheme 2.14**) can be carried out through condensation of an aldehyde with an excess of pyrrole (1:25 to 1:100 ratios) in order to suppress the formation of oligomers beyond the dipyrromethane stage. Typically, this reaction is carried out at room temperature and requires the addition of TFA or Lewis acids, such as BF_3 or $InCl_3$, to activate the aldehyde. After the reaction is complete, the excess of pyrrole can be recovered by vacuum distillation, and the product purified by column chromatography (small-scale) or by sublimation (gram-scale). It is worth mentioning that

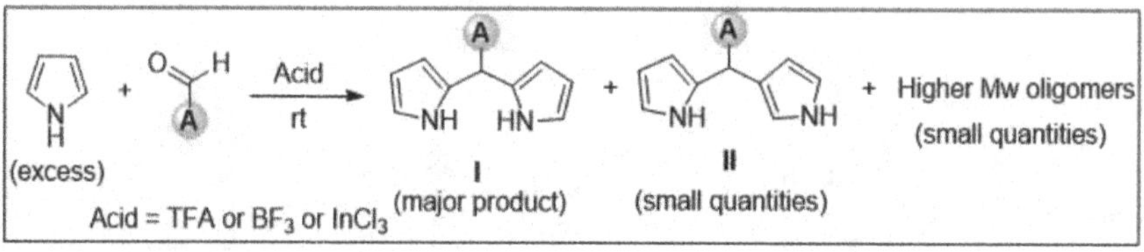

SCHEME 2.14 Typical one-pot synthesis of dipyrromethane.

small quantities of the so-called "N-confused" dipyrromethane **II** can be formed as a result of an attack of one pyrrole through its 3 position. To avoid the formation of such side products, more laborious multi-step procedures can be employed [92,93], but these are outside the scope of this chapter.

Starting from dipyrromethane **I**, a more complex structure can be obtained, such as the 1,9-diacyl and 1,9-dicarbinol dipyrromethanes (**Scheme 2.15a**). Symmetrical 1,9-diacyl dipyrromethanes **III** can be obtained through a conventional Friedel–Crafts acylation, using a Lewis acid ($SnCl_4$ or $SbCl_5$) [94] and two molar equivalents of an acyl chloride containing the desired group **B**. The mechanism (**Scheme 2.15b**) comprises the activation of the acyl chloride with the Lewis acid ($SnCl_4$ was used as an example), through the nucleophilic attack of the chlorine to tin, with consequent formation of a Lewis adduct. Then, the chlorine is removed from the molecule, leading to the formation of an acyl cation (stabilized by two resonance structures). Finally, the nucleophilic attack of the 2 position of pyrrole to the acyl cation, followed by regeneration of pyrrole, gives the 1-acyl dipyrromethane. The synthetic route proceeds through the reaction with another molar equivalent of the acyl chloride, giving the 1,9-diacyl dipyrromethane **III**.

The reduction of both carbonyl groups of **III** to the corresponding alcohols yields the highly reactive 1,9-dicarbinol dipyrromethane **IV** (**Scheme 2.15a**). This reduction step can be carried out by using $NaBH_4$ (**Scheme 2.15c**). It encompasses a hydride transfer from $NaBH_4$ to the carbon of the carbonyl group, forming an alkoxide intermediate. In the presence of a proton source (e.g., water), the alcohol is formed. This reaction is repeated for the other acyl group, yielding the 1,9-dicarbinol dipyrromethane **IV**. The synthesis of unsymmetrical 1,9-diacyl dipyrromethanes can also be carried out (**Scheme 2.15a**) [95]. This is possible through the treatment of dipyrromethane **I** with EtMgBr (2 to 2.5 mol equivalents), which deprotonates pyrrole's N-H groups (**Scheme 2.15d**), facilitating pyrrole's 2

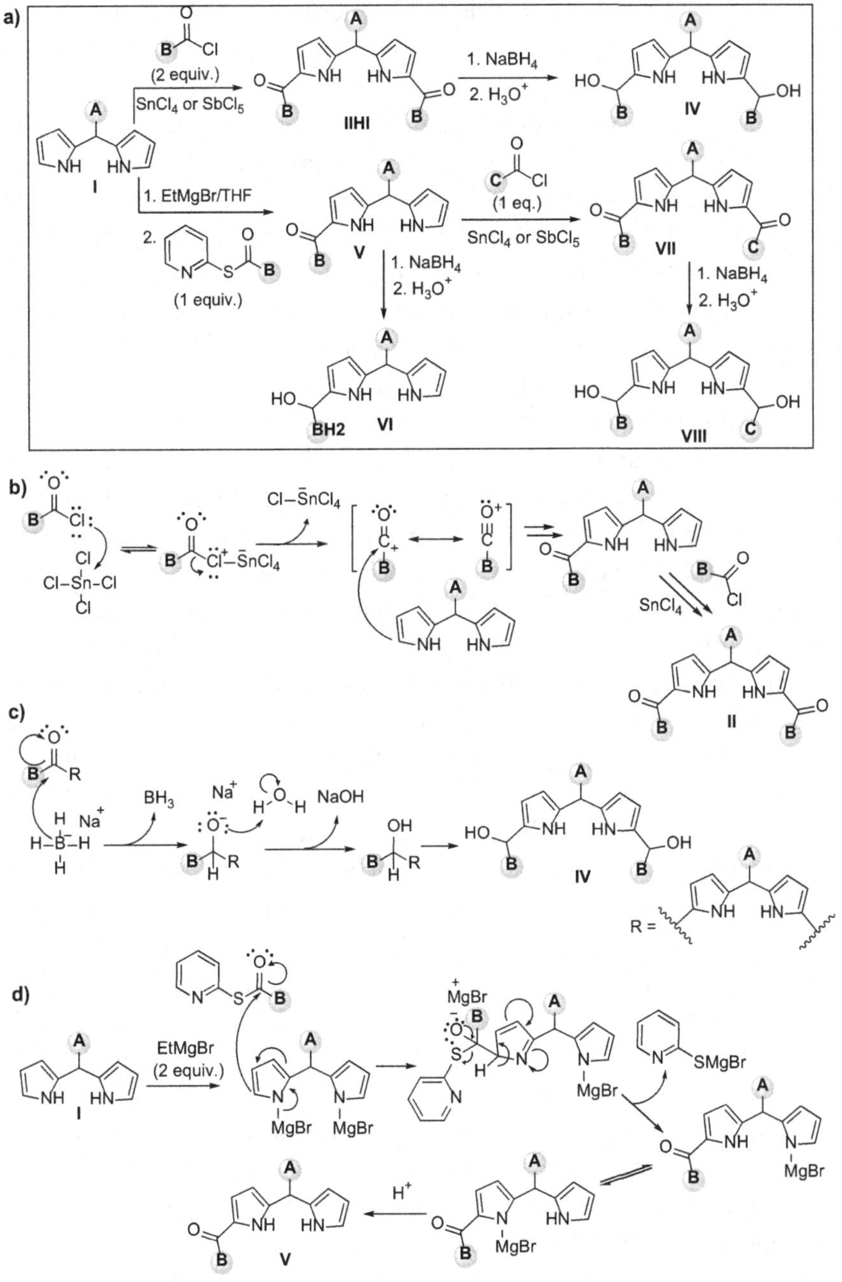

SCHEME 2.15 (a) Generic synthesis of 1,9-diacyl and 1,9-dicarbinol dipyrromethanes; (b) mechanism for the Friedel–Crafts acylation of dipyrromethane; (c) mechanism for the reduction of acyl dipyrromethanes to their carbinol derivatives; (d) mechanism for the selective mono-acylation of dipyrromethane using a Mukaiyama reagent.

position nucleophilic attack. When this intermediate is treated with one molar equivalent of a Mukaiyama reagent containing the desired group **B** (prepared through reaction of 2-mercaptopyridine with an acyl chloride containing **B**), a nucleophilic attack of pyrrole to the carbonyl carbon occurs. The tetrahedral intermediate formed proceeds with the regeneration of the carbonyl group, with 2-mercaptopyridine serving as leaving group. Meanwhile, pyrrole aromaticity is regenerated through elimination of H^+, as typical in these reactions. This will lead to a protonation equilibrium of the pyrrolic NH groups, which will hinder further acylation reactions, and explains the observed selectivity for mono-acylation. Finally, acidic work-up yields the 1-acyldipyrromethane **V**.

Following the aforementioned mechanisms, **V** can then react with $NaBH_4$ to yield 1-carbinoldipyrromethane **VI**. Alternatively, it can react with an acyl chloride containing group **C**, yielding the 1,9 unsymmetrical 1,9-diacyl dipyrromethane **VII** and then the corresponding 1,9-dicarbinol dipyrromethane **VIII** after reduction with $NaBH_4$ (**Scheme 2.15a**).

The preceding plethora of dipyrromethane derivatives can then be used for the synthesis of *trans*-A_2B_2 porphyrins through many different routes, as exemplified in **Scheme 2.16**. Through a two-step synthetic methodology, the base dipyrromethane **I** can react with an aldehyde in the presence of an acid (TFA or Lewis acid), allowing the formation of the porphyrinogen, which is oxidized in the second step with DDQ or *p*-chloranil (***Via* A**). Alternatively, the 1,9-dicarbinol dipyrromethane **IV** can be used as condensation partner, instead of the aldehyde (***Via* B**) [96]. When using 1-carbinol dipyrromethane **VI**, the synthesis proceeds through a self-condensation approach (***Via* C**) [95]. These last two approaches can proceed using milder Lewis acid catalysts (e.g. $Yb(OTf)_3$, $Dy(OTf)_3$, $Sc(OTf)_3$, $InCl_3$), as they employ the more reactive carbinol dipyrromethanes, with the advantage of reducing porphyrin scrambling. It is worth reminding the reversible nature of all the synthetic steps leading up to the porphyrinogen formation (see **Scheme 2.2**), where strong acids can promote breaking of the bond between pyrrole and their methane bridge, with consequent recombination of these fragments into linear polymers or porphyrinogens with unexpected substitution patterns.

Another remarkable approach to *trans*-A_2B_2 porphyrins is the self-condensation of 1-acyl dipyrromethane **V**, but under basic conditions, using 1,8-diazabicyclo(5.4.0)undec-7-ene (DBU) as base, magnesium (II) as metal template for the cyclization of the dipyrromethanes, and air as oxidant [97]. This methodology, reminiscent of the original *Rothemund*

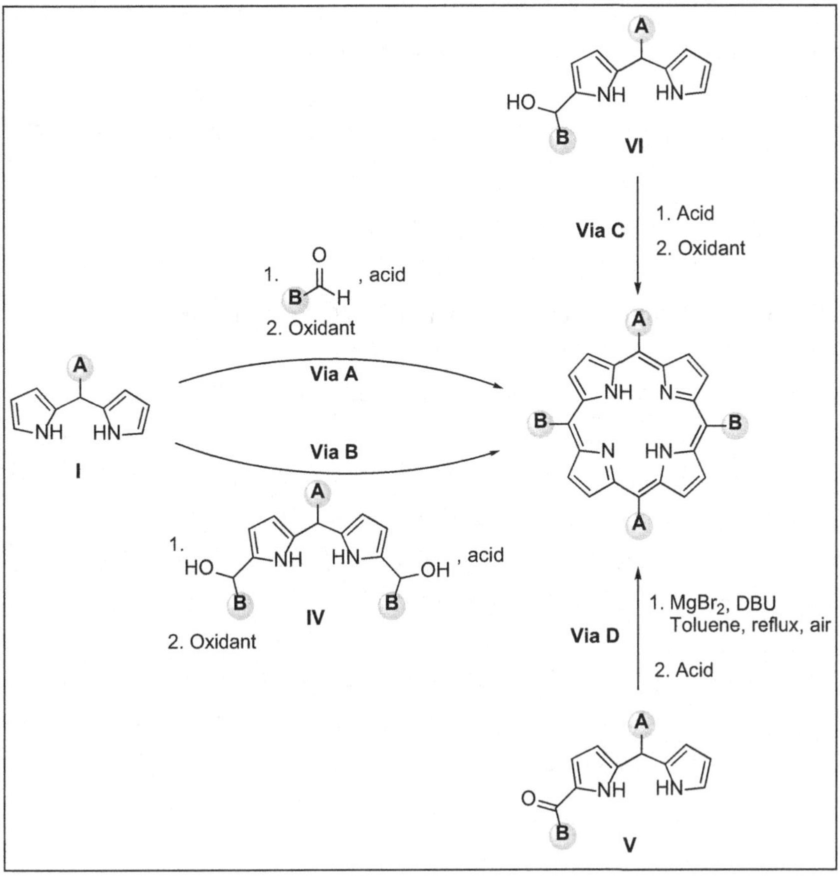

SCHEME 2.16 Main synthetic routes for obtaining *trans*-A_2B_2 porphyrins from dipyrromethanes and their corresponding carbinol derivatives.

approach, yields the Mg(II) metalloporphyrin complex, which can be treated with mild acids to obtain the free base porphyrin.

Besides *trans*-A_2B_2 porphyrins, other porphyrinic macrocycles with more complex substitution patterns in their *meso* positions can also be prepared, including ABCD porphyrins **(Scheme 2.17)** [98,99]. Having in hand a non-symmetrical 1,9-dicarbinol dipyrromethane **VIII** with groups A, B, and C, a dipyrromethane containing group D can be used as condensation partner, under conditions described in **Scheme 2.16**, yielding the desired ABCD porphyrin.

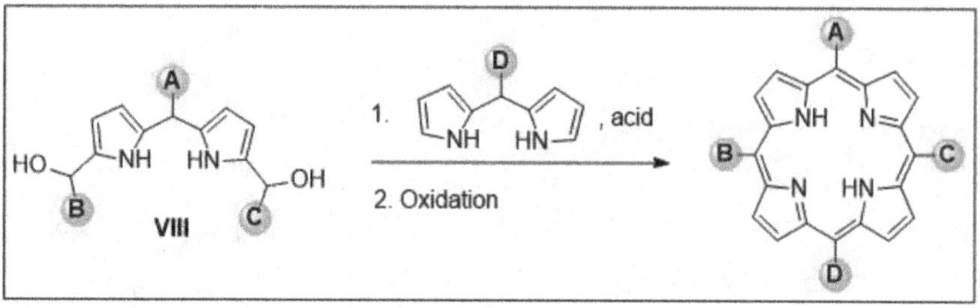

SCHEME 2.17 Synthetic route for obtaining ABCD porphyrins.

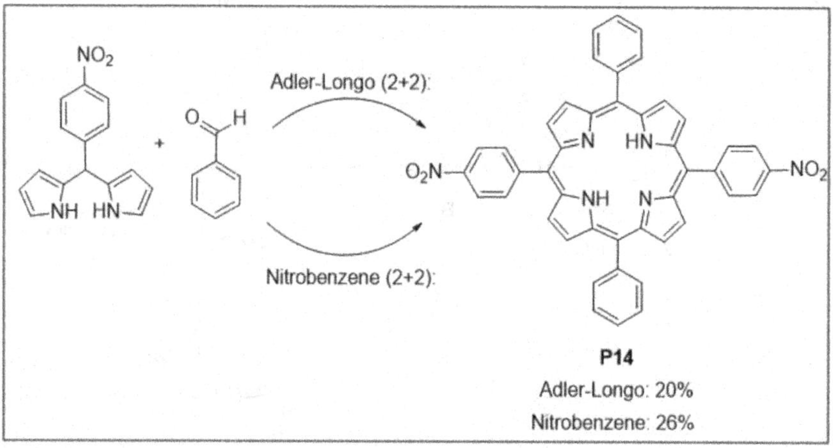

SCHEME 2.18 Synthesis of non-symmetrical **P14** through dipyrromethane-based approaches.

In **Scheme 2.18**, one example of a non-symmetric porphyrin, **P14**, synthesized through dipyrromethane-based approaches, is presented [100,101].

The synthesis of **P14**, previously mentioned in **Table 2.4**, can also be achieved by a 2 + 2 condensation reaction, using 5-(4-nitrophenyl)dipyrromethane (obtained in 58%, following the general procedure described in **Scheme 2.11**) and benzaldehyde as reagents. Both were based in one-pot methodologies, *Adler–Longo* (20% yield) [100] and *nitrobenzene* (26% yield) [101], which represent better yields when compared with the aldehyde mixture approach (12%, **Table 2.4**). However, while both dipyrromethane-based methodologies offer better yields in the porphyrin synthesis step, it should be noted that, when also considering the dipyrromethane synthesis step, the overall yield decreases to 12 and 15% for the *Adler–Longo* and *nitrobenzene* methods, respectively. The reduction of the nitro groups in **P14** for bioconjugation will be discussed in Section 3.5a.

Based on the aforementioned techniques for the synthesis of non-symmetrical porphyrins, *Lindsey* described a non-symmetrical chlorin containing alkyne groups for further bioconjugation with N_3-containing PEG groups, through click chemistry (**C3; Scheme 2.19**) [102].

Its synthesis starts with the tri-alkylation of the 2,4,6-trihydroxybenzaldehyde with an alkyne-containing alkyl bromide through a conventional S_N2 reaction. Then, this alkyne-functionalized aldehyde reacts with an excess of pyrrole and $InCl_3$ to form the corresponding dipyrromethane, as previously described in **Scheme 2.15** [103]. Then, taking advantage of the high reactivity of pyrrole's 2 position, it was possible to achieve a

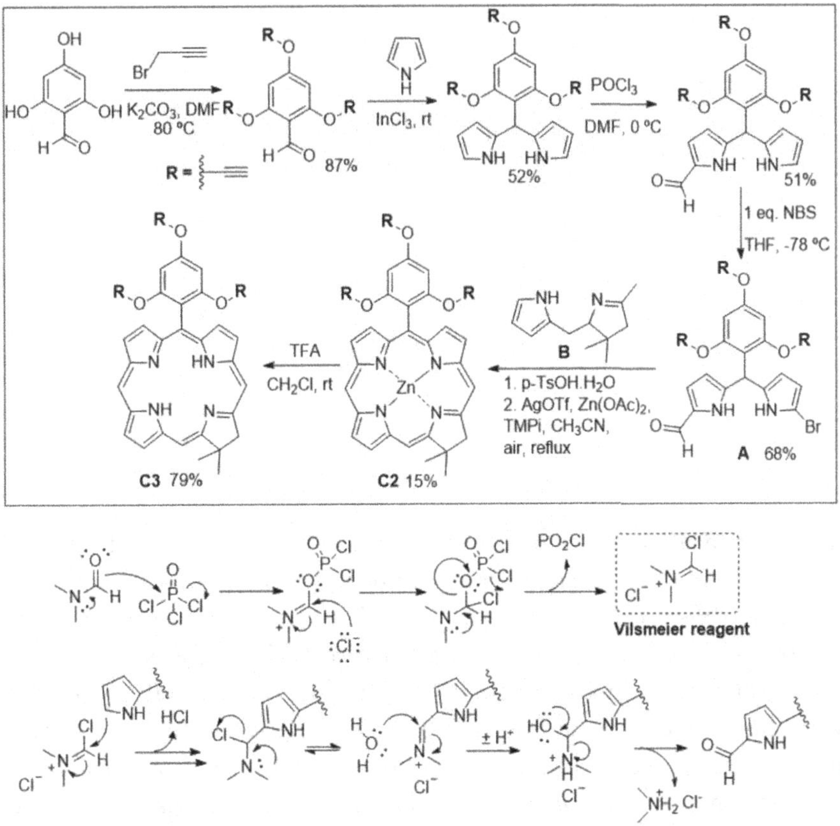

SCHEME 2.19 Multi-step synthesis of non-symmetrical **C2**.

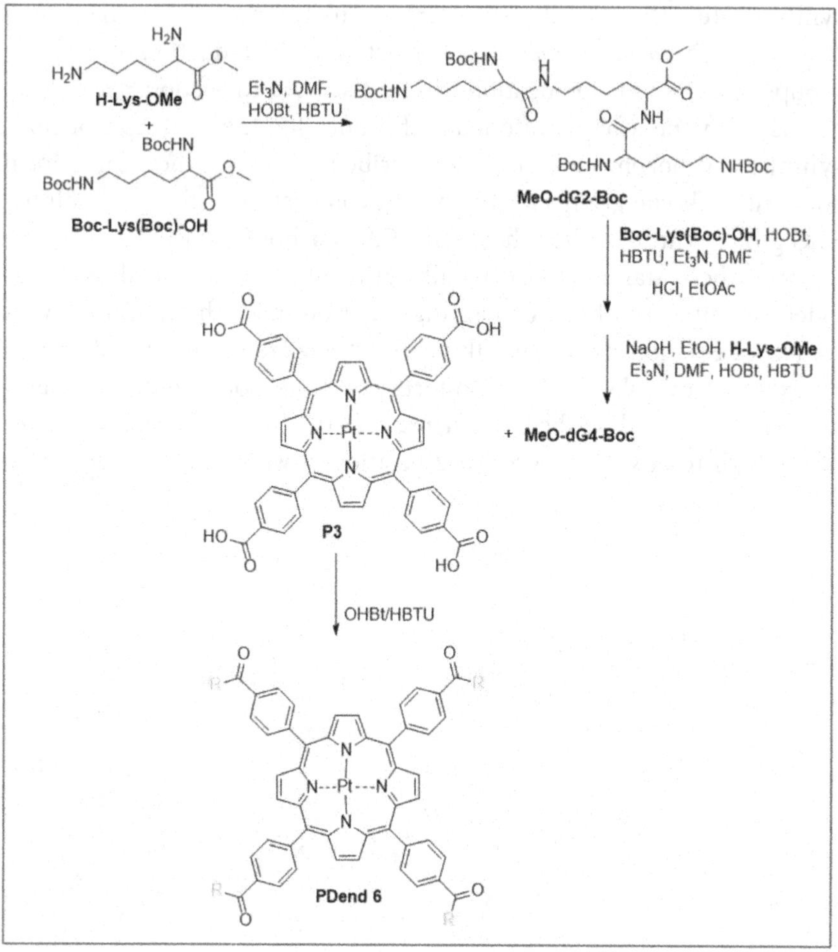

SCHEME 2.19 (Continued)

mono-formylation of the dipyrromethane, with some degree of selectivity (51% yield of the desired 2-formyl product, compared with only 20% of the 2,9-diformyl biproduct), through a Vilsmeier–Haack reaction. This reaction proceeds with the reaction between DMF and $POCl_3$, forming the Vilsmeier reagent **(Scheme 2.19)**. This electrophile will then react with dipyrromethanes at their more reactive 2 position through a standard aromatic electrophilic substitution, with elimination of HCl. Then, an iminium cation is formed, followed by nucleophilic attack of water and elimination of dimethylammonium chloride, resulting in the formation of the 2-formyldipyrromethane.

After the 2-formylation of the substituted dipyrromethane, its 9 position is brominated by the addition of 1 equivalent of NBS, through a standard electrophilic aromatic substitution, forming a 2-formyl-9-bromo-dipyrromethane derivative (**Scheme 2.19, A**). This dipyrromethane is then combined with a reduced derivative of another dipyrromethane (**Scheme 2.19, B**; see reference [104] for details on its synthesis) and, through a two-step procedure, forms the Zn(II) chlorin **C2** in 15% yield. In this reaction, the first step refers to the *p*-toluenesulfonic acid–catalyzed condensation between intermediate **B** and the formyl group in **A**. Then, ring closing and oxidation (second step) are achieved by atmospheric oxygen and the addition of AgOTf, $Zn(OAc)_2$ (to facilitate chlorin cyclization through template effect), and TMPi (neutralization of the acid) [105]. Finally, TFA-promoted Zn(II) removal from the porphyrin core allows preparation of the free base **C3** in 79% yield [103].

It is clear that the chemical versatility of pyrrole and dipyrromethanes opens the way for the synthesis of non-symmetric *meso*-substituted porphyrins with a great variety of substituents and substitution patterns. The methodologies presented in this subchapter are but few examples of the approaches reported over the years.

Another area of great interest in the preparation of bioconjugates relates to the post-cyclization synthetic modification of porphyrins for introduction of appropriate functional groups for linking to the desired biomolecule. This will be presented and discussed in the next section.

2.5 PORPHYRIN STRUCTURAL MODIFICATIONS

2.5.1 Nitration and Reduction to Amines

In the preparation of bioconjugates, nitration is one of the most important reactions for porphyrin post-cyclization modifications. Over the years, many synthetic methodologies have been developed for the selective introduction of nitro groups at porphyrin's *meso*- and β-positions [106]. In tetra-*meso*-substituted porphyrins, mono-nitration at the β-positions can be achieved by reaction of the free base porphyrins or metalloporphyrins with nitric acid (**Scheme 2.20a**). When using free base porphyrins as substrates, fuming nitric acid is required, and relatively low yields (up to 40%) of the desired mono-β-nitro porphyrin are achieved, with frequent contamination with *meso*-aryl ring nitration secondary products [106]. On the other hand, nitration of Cu(II), Ni(II), or Pd(II) metalloporphyrins proceeds much smoother over the corresponding free base porphyrins [107]. For instance, the synthesis of a Cu(II) metalloporphyrin can be easily achieved by reaction of the free base porphyrin with $Cu(OAc)_2$, in DMF, at reflux temperature (**Scheme 2.20a, *via* A**). Then, the porphyrin is dissolved in a chlorinated solvent and reacted with 25% HNO_3, at room temperature, yielding the desired mono-β-nitro porphyrin, in yields up to 77%, with small contaminations with the corresponding di-β-nitro porphyrin.

An alternative one-step approach consists in the use of $Cu(NO_3)_2$ and acetic anhydride, which concomitantly promotes the formation of the Cu(II) porphyrin complex and the mono-β-nitration, yielding the mono-β-nitro Cu(II) porphyrin in yields up to 86% (**Scheme 2.20a, *via B***) [108]. The corresponding free base porphyrin can then be obtained in 76–94% yields by reacting the Cu(II) with a strong acid, such as TFA or H_2SO_4 [108].

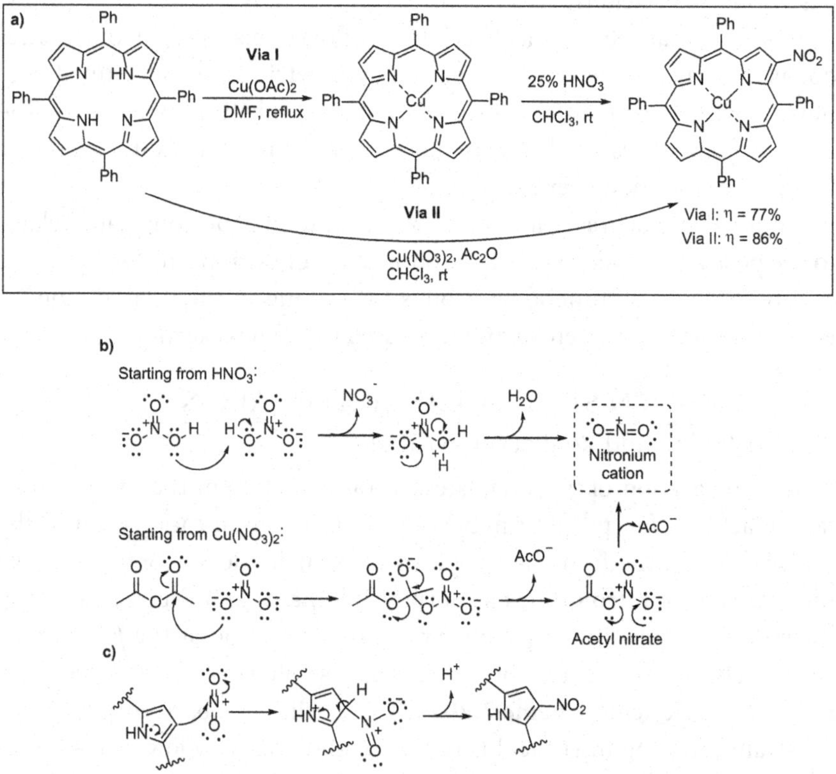

SCHEME 2.20 (a) Synthetic strategies for selective mono-β-nitration of porphyrins; (b) mechanisms for the generation of the nitronium cation, starting from either HNO_3 or $Cu(NO_3)_2$; (c) general mechanism for the nitration of an aromatic ring with the nitronium cation.

The general mechanisms involved in the nitration of porphyrins, using either HNO_3 or $Cu(NO_3)_2$, are depicted in **Scheme 2.20b**. When HNO_3 is used as starting reagent (i.e., ***via* A** in **Scheme 2.20a**), it can be protonated by another HNO_3 molecule, which promotes dehydration and generation of the desired nitronium cation. On the other hand, when $Cu(NO_3)_2$ is used in combination with acetic anhydride (Menke nitration) [109], there is a nucleophilic attack of the nitrate ion to acetic anhydride, followed by elimination of acetate and formation of the intermediate acetyl nitrate. Then, elimination of another acetate ion leads to the formation of the nitronium cation (**Scheme 2.20b**). This cation will act as the electrophile in an electrophilic aromatic substitution reaction (**Scheme 2.20b**), which proceeds through the nucleophilic attack of the aromatic ring, with formation of a cationic non-aromatic intermediate (but stabilized through resonance by the ring's π system). The reaction proceeds by proton elimination with consequent regain of the ring aromaticity. In the case of porphyrins, the extensive aromatic resonance structure greatly helps in the stabilization of this reaction's intermediate, which explains why porphyrin β-nitration can proceed in such mild conditions. β-nitro porphyrins can be further modified in order to allow effective reactions with electrophiles (**Scheme 2.21**). The first step consists in the reduction of the nitro group to the corresponding amine, through standard methodologies. This reaction proceeds smoother when using metalloporphyrins containing the aforementioned β-activating metals, such as Cu(II), and yields a more stable product.

The general mechanism for the reduction of nitroaromatics to their corresponding amines using Sn/HCl is described in **Scheme 2.21**. It entails the coordination of the Sn metal with the nitro group, followed by acid-promoted dehydration and elimination of $SnCl_2$, forming the nitroso (R-N=O) intermediate. Then, another reduction step occurs with the formation of $SnCl_2$, resulting in the *N*-hydroxylamine (R-NH-OH) intermediate. A final Sn-mediated 2-electron reduction occurs, where the amine group is formed with a concomitant dehydration and formation of $SnCl_2$.

In particular, when carrying out the reduction of mono-β-nitro Cu(II) TPP complex using metallic Sn and concentrated HCl under ultrasonic irradiation, the corresponding mono-β-amino Cu(II) TPP complex can be obtained in yields up to 80% [108]. This β-amino group can then react with linkers containing appropriate electrophilic (**E**) groups, namely, through S_N reactions or addition to activated carbonyl groups (**Scheme 2.21**).

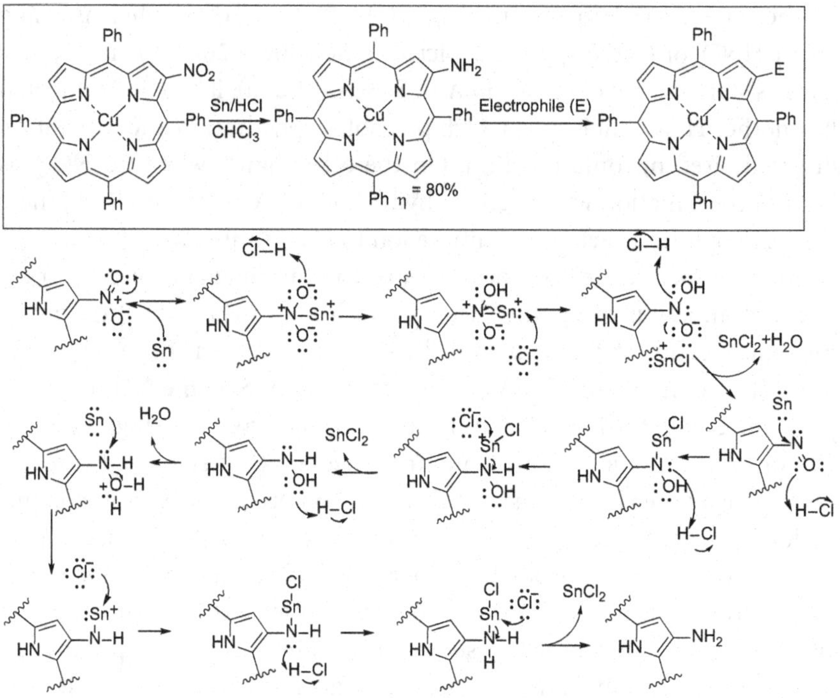

SCHEME 2.21 Modulation of β-nitro porphyrins for reaction with nucleophiles and electrophiles.

Besides the β-pyrrolic position, nitration in the *meso*-aryl positions can also be achieved under different conditions. For example, *Vicente* reported a procedure where **TPP** reacted with $NaNO_2$, in TFA, at room temperature, for just 90 sec (**Scheme 2.22**) [110].

TFA can protonate the macrocycle inner nitrogen, reducing the reactivity of the β-pyrrolic positions and favoring nitration at the aryl substituents. The selectivity for mono- and di-nitration can be controlled by using an appropriate number of equivalents of $NaNO_2$. For instance, when promoting the di-nitration, 1.8 equivalents of $NaNO_2$ were used, affording the 5,15-bis(4-nitrophenyl)-10,20-diphenylporphyrin **P14** and its isomer, 5,15-bis(4-nitrophenyl)-10,20-diphenylporphyrin, as main products. The $NaNO_2$/TFA method is yet another nitration strategy that, through a series of sequential reactions (**Scheme 2.22**), culminates in the formation of the nitronium cation. As shown earlier (**Scheme 2.20**), this nitronium cation will act as an electrophile and react with the aromatic ring, leading to the formation of the desired nitroaryl compounds. Without isolation of

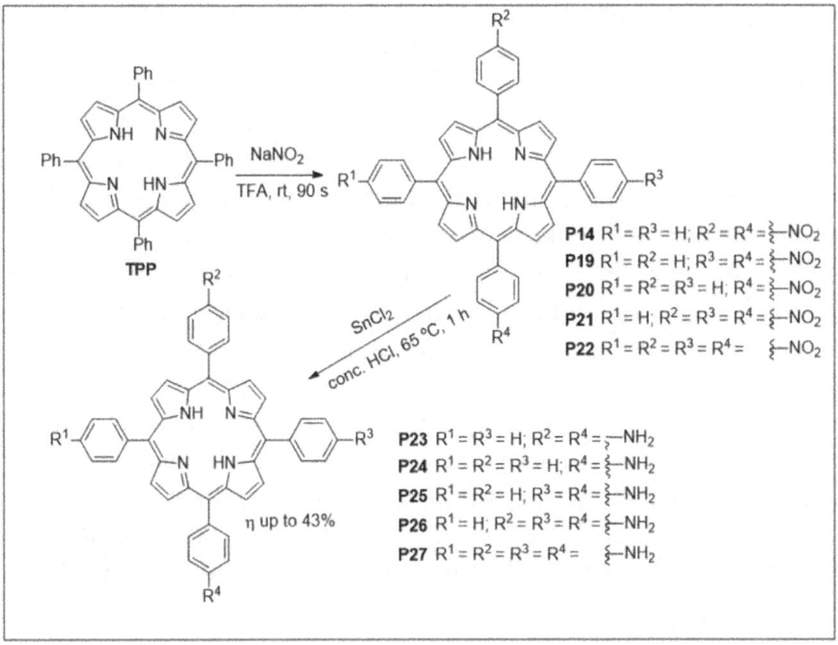

NaNO$_2$, TFA, rt, 90 s

P14 $R^1 = R^3 = H$; $R^2 = R^4 = -NO_2$
P19 $R^1 = R^2 = H$; $R^3 = R^4 = -NO_2$
P20 $R^1 = R^2 = R^3 = H$; $R^4 = -NO_2$
P21 $R^1 = H$; $R^2 = R^3 = R^4 = -NO_2$
P22 $R^1 = R^2 = R^3 = R^4 = -NO_2$

SnCl$_2$, conc. HCl, 65 °C, 1 h

P23 $R^1 = R^3 = H$; $R^2 = R^4 = -NH_2$
P24 $R^1 = R^2 = R^3 = H$; $R^4 = -NH_2$
P25 $R^1 = R^2 = H$; $R^3 = R^4 = -NH_2$
P26 $R^1 = H$; $R^2 = R^3 = R^4 = -NH_2$
P27 $R^1 = R^2 = R^3 = R^4 = -NH_2$

η up to 43%

$$NaNO_2 + CF_3CO_2H \rightleftharpoons HNO_2 + CF_3CO_2Na$$
$$2HNO_2 \rightleftharpoons N_2O_3 + H_2O$$
$$HNO_2 \overset{H^+}{\rightleftharpoons} NO^+ + H_2O$$
$$N_2O_3 \rightleftharpoons NO + NO_2$$
$$N_2O_3 \rightleftharpoons NO^+ + NO_2^-$$
$$2NO_2 \rightleftharpoons N_2O_4 \overset{H^+}{\rightleftharpoons} NO_2^+ + HNO_2$$

SCHEME 2.22 Synthesis of amino-phenyl porphyrins **P14**, **P19**, to **P20–P27** through di-nitration of **TPP**, followed by nitro reduction with $SnCl_2$/HCl.

P14 and its 5,10-isomer, *Vicente* proceeded with the reduction of the nitro groups with $SnCl_2$, in concentrated HCl (see mechanism in **Scheme 2.21**), for 65°C in 1 h. The desired 5,15-bis(4-aminophenyl)-10,20-diphenylporphyrin (**P23**) was separated from its isomer by silica gel chromatography and isolated with an overall yield of 21%.

By changing the number of equivalents of $NaNO_2$ for TPP nitration, *Vicente* [111] also reported mono-, di-, tri-, and tetra-nitrophenyl porphyrins (**P19** to **P22**; **Scheme 2.22**). Then, after $SnCl_2$/HCl reduction, the corresponding **P24** and **P27** were obtained. These aminophenyl groups were then treated with glycolic anhydride, in DMF, at room temperature, for 16 h, affording the corresponding carboxylic acid–functionalized porphyrins (**P28** to **P31**) in 95 to 100% yields (**Scheme 2.23**). The conjugation of these compounds with amino-functionalized PEG linkers will be described in **Chapter 4**.

Based on the TFA/$NaNO_2$ *meso*-phenyl nitration strategy, *Malinowski* synthesized a set of mono-nitrated porphyrins, starting from **P10** and **P11** (**Scheme 2.24**) [78]. Unlike previous methodologies described in this section, the nitro group in both porphyrins serves only as an activating group for the S_NAr reaction, where a hydroxyl-containing sugar (**Chapter 5**) will react with the carbon *ipso* to the halogen, with consequent formation of a new C-O bond.

Overall, nitration of *meso*-aryl porphyrins confers two main pathways for bioconjugation: (a) appropriate conditions that favor mono-functionalization at the *β*-positions, which in turn allows selectivity for mono-conjugation with the desired biomolecule; (b) functionalization at 1, 2, 3, or positions in the *meso*-aryl groups, allowing conjugation with multiple biomolecules.

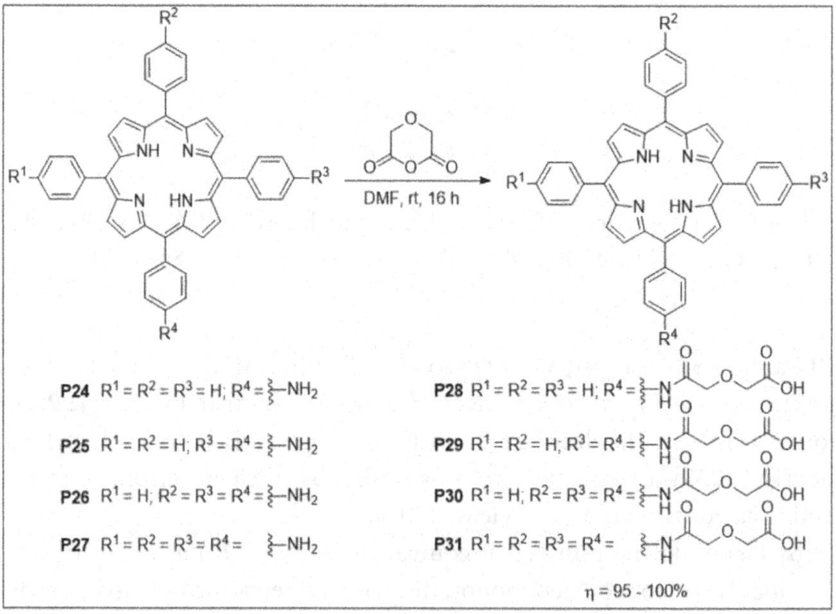

SCHEME 2.23 Synthesis of carboxylic acid–functionalized porphyrins (**P28** to **P31**) by reaction with glycolic anhydride.

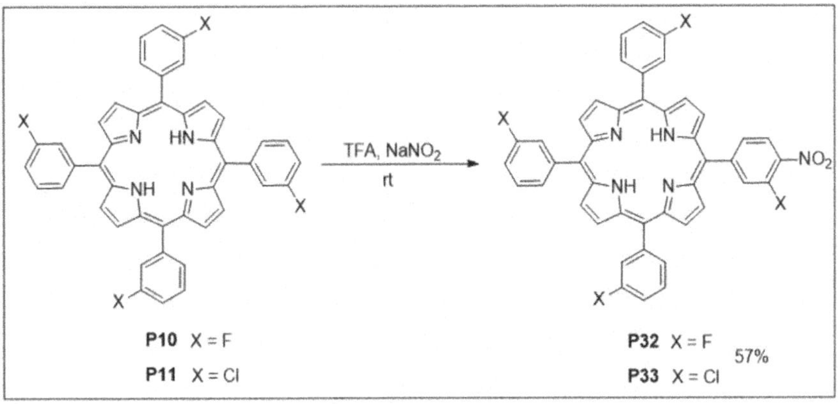

SCHEME 2.24 Mono-nitration of *meso*-3-fluoro and 3-chlorophenyl porphyrins.

2.5.2 Formation of Diazonium Salts

Besides their direct use as nucleophiles for the preparation of bioconjugates, amino-aryl groups embedded in porphyrins can be converted into diazonium salts, which are excellent leaving groups for S_NAr reactions. *Reiser* [112] used this strategy to prepare 5,10,15-tris(4-(*tert*-butyl)phenyl)-20-(4-azidophenyl)porphyrin **P35** (**Scheme 2.25**) for use in bioconjugation through click chemistry. Its synthetic route starts with the synthesis of 5,10,15-tris(4-(*tert*-butyl)phenyl)-20-(4-nitrophenyl)porphyrin **P15** (previously described in **Table 2.4**, entry 3), followed by a reduction using the $SnCl_2/HCl$, affording the corresponding **P34** in 76% yield. Then, by treating **P34** with $NaNO_2$ and TFA, at 0°C, followed by the addition of NaN_3, the aryl azide–containing **P35** is obtained in 95% yield. The mechanism for this two-step reaction (**Scheme 2.25**, step 1) starts with the TFA-mediated protonation of the nitrite ion (NO_2^-) to form nitrous acid (HNO_2). A second protonation step, followed by dehydration, yields the nitrosonium ion, which is the electrophile in this diazotization reaction. Its reaction with the amine group in the phenyl ring forms an unstable *N*-nitrosoaminium ion as an intermediate which, upon deprotonation, forms an *N*-nitrosoamine. This intermediate is converted to a diazohydroxide, which will undergo dehydration in the presence of an acid to form the diazonium ion. The second step involves the reaction of the diazonium salt with NaN_3, whose mechanism (**Scheme 2.25**, step 2) is thought to proceed via an acyclic zwitterionic intermediate, which then forms the aryl azide by elimination of N_2 [113].

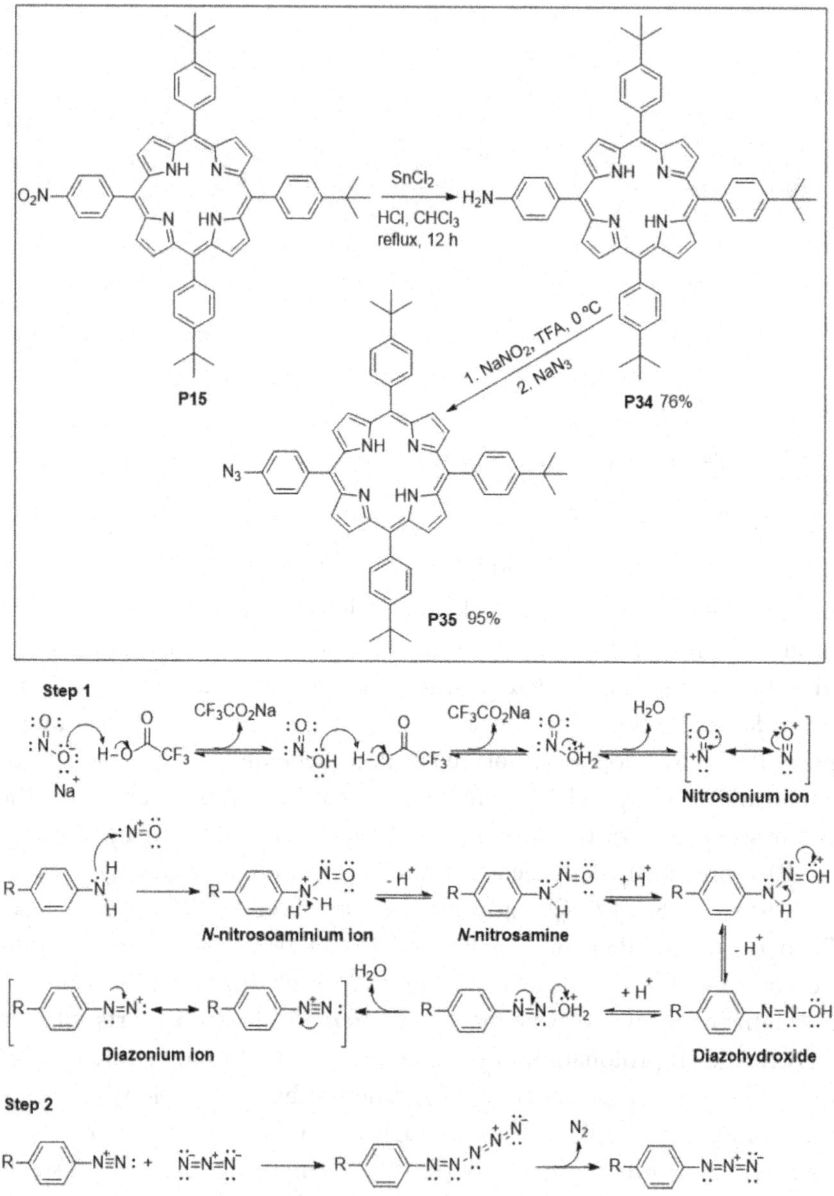

SCHEME 2.25 Synthesis of **P35** containing an azide group for further conjugation via click chemistry.

2.5.3 Sulfonation and Chlorosulfonation

Sulfonation and chlorosulfonation are other typical reactions for the functionalization of aromatic groups, which can be used for the preparation of bioconjugates with *meso*-aryl porphyrins (**Scheme 2.26**).

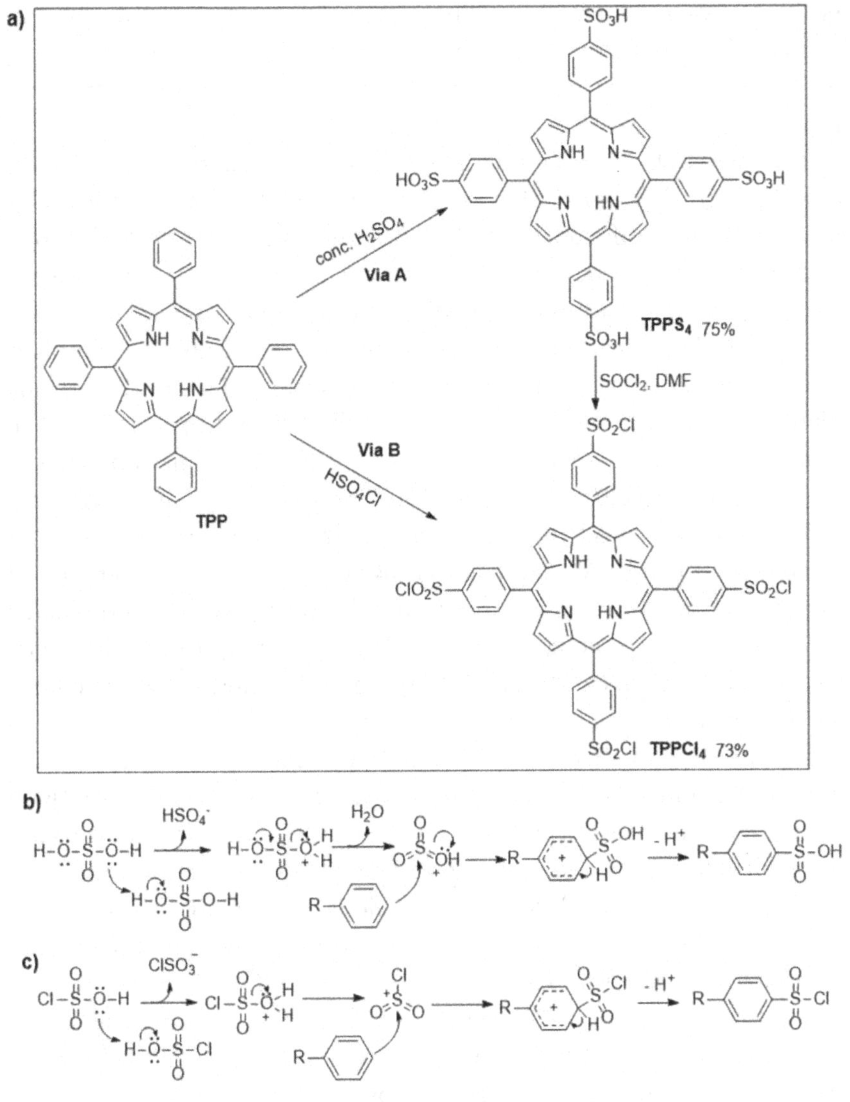

SCHEME 2.26 (a) Strategies for obtaining 5,10,15,20-tetrakis(4-chlorosulfonyl-phenyl)porphyrin ($TPPCl_4$); (b) mechanism for aromatic sulfonation; (c) mechanism for aromatic chlorosulfonation.

In particular, chlorosulfonic groups are of great interest due to their high reactivity with nucleophiles, affording many possibilities for bioconjugation. These groups can be introduced using concentrated sulfuric acid (**via A**, two-step methodology, **Scheme 2.26a**) or chlorosulfonic acid (**via B**, one-step methodology, **Scheme 2.26a**), and in both cases, selectivity towards functionalization of the meso-aryl substituents over β-pyrrolic positions can be achieved. Since both are relatively strong acids, protonation of the macrocycle's inner nitrogen atoms occurs, strongly deactivating the β-pyrrolic positions towards these types of aromatic electrophilic reactions.

The preparation of 5,10,15,20-tetrakis(4-sulfonatophenyl)porphyrin (**TPPS$_4$**) directly from **TPP** (Via A) can be achieved by dissolving **TPP** in concentrated sulfonic acid and leaving the reaction at 100°C for 16 h. One of the major issues with this synthetic approach is the purification of **TPPS$_4$**, which requires precipitation of the product and neutralization using a sodium bicarbonate aqueous solution. Since this porphyrin is water-soluble, the inorganic salts need to be separated from **TPPS$_4$** using dialysis membranes with small pores that only allow the passage of small molecules and thus can selectively retain **TPPS$_4$**. After this cumbersome purification step, **TPPS$_4$** is obtained in 75% yield [114]. The classic aromatic sulfonation mechanism (**Scheme 2.26b**) entails the intermolecular protonation of sulfuric acid, followed by dehydration and generation of sulfur trioxide, which exists in an equilibrium between its protonated (SO_3H^+) and non-protonated forms (SO_3). This strong electrophile then participates in a typical aromatic electrophilic substitution, which culminates in the formation of the sulfonic group.

The **TPPS$_4$** can then be converted to the corresponding 5,10,15,20-tetrakis(4-chlorosulfonylphenyl)porphyrin (**TPPCl$_4$**) by reaction with thionyl chloride ($SOCl_2$), in DMF, for 3 h at 50°C. This product is water-insoluble and thus can be easily precipitated with water from the reaction mixture and filtered. No isolated yield is reported for this step, since **TPPCl$_4$** can be used immediately in reactions with nucleophiles, in order to avoid hydrolysis of the chlorosulfonic groups [115].

Alternatively, a convenient one-step synthetic approach (**Scheme 2.26a**, via B) for obtaining **TPPCl$_4$** can be achieved by treating **TPP** with chlorosulfonic acid, at room temperature, for just 1 h. The work-up is much simpler than via A since, after reaction completion, chloroform and water can be added, forming a biphasic system where **TPPCl$_4$** remains in the organic phase and the excess of acid can be removed by multiple washings with water and a sodium bicarbonate solution. After solvent evaporation, **TPPCl$_4$** can be isolated in 73% yield [116]. The mechanism for chlorosulfonation (**Scheme 2.26c**) [117] differs from the classic sulfonation in the generation of the electrophile. Here, an intermolecular proton transfer between two sulfonic acid molecules leads to dehydration and formation of the active electrophile, which in this case is ClO_2S^+.

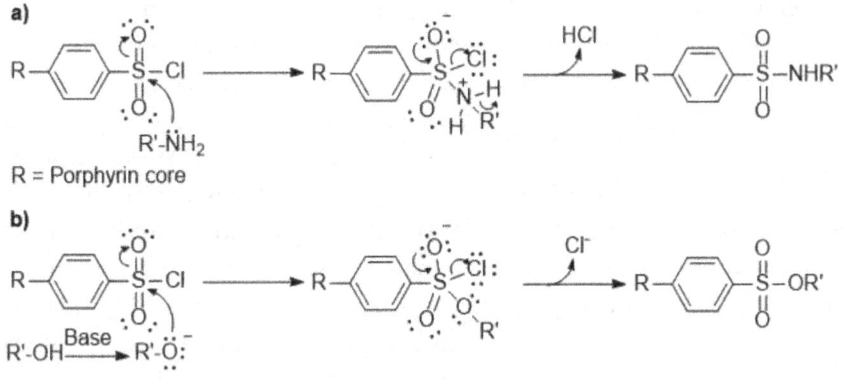

SCHEME 2.27 Mechanism for the synthesis of (a) sulfonamides and (b) sulfonate esters, starting from the sulfonyl chlorides contained in **TPPCl$_4$**.

While only **TPP** is given here as a model substrate for chlorosulfonation, this reaction can even be applied to *meso*-aryl porphyrins containing electron-withdrawing halogen substituents, such as chlorine and fluorine atoms, requiring just an increase in reaction temperature (to 50–100°C) and/or reaction time (1.5–3 h) [118].

As previously mentioned, **TPPCl$_4$** is an attractive intermediate for bioconjugation since it can react with either amines, forming sulfonamides [119] (**Scheme 2.27a**), or alcohols, originating sulfonate esters [120] (**Scheme 2.27b**).

Typically, both reactions can be carried out at room temperature in 1–3 h [119,120], although the formation of sulfonate esters requires the deprotonation of the alcohol by a relatively stronger base, which may constitute a disadvantage if the bioconjugate contains sensitive functional groups.

2.5.4 Bromination

Bromo-aryl compounds have enjoyed great success as substrates for Pd-catalyzed coupling reactions, and thus, it is of great interest to develop methods to selectively introduce bromine atoms in porphyrin rings and take advantage of transition metal catalysis for the synthesis of bioconjugates.

In the case of 5,15-diphenylporphyrin (**DPP**), its reaction with two molar equivalents of *N*-bromosuccinimide (NBS), in a chlorinated solvent such as chloroform, and at 0°C to room temperature, the di-brominated derivative **P36** can be obtained in very high yields (up to 95%;

Scheme 2.28) [121]. Thus, with appropriate reaction conditions (i.e., without excess of NBS and low temperatures), bromination at porphyrin's free *meso* positions can be selectively achieved over their β-pyrrolic positions. By reacting **DPP** with just 0.8 equivalents of NBS, the mono-brominated derivate **P37** can be obtained in yields up to 66% [122]. However, some contamination with **P36** is expected, as well the presence of unreacted **DPP**, and thus, **P37** isolation is more challenging due to the similar polarities of brominated and non-brominated porphyrins. This issue can be overcome by employing more complex bromination methodologies, such as using organopalladium intermediates [123]. If no free *meso* positions are available, bromination can occur in the β-pyrrolic positions [124].

Using a chlorin derivative (**C3**) whose synthesis was previously described in Section 2.4 (**Scheme 2.19**), *Lindsey's* group achieved a *meso* mono-bromination by using 1.1 equivalents of NBS, in THF, at room temperature (**Scheme 2.29**). The desired mono-brominated product **C4** was obtained in 87% isolated yield, with minimal formation of side products (β-mono-bromination: 3%; di-bromination: 7%) [102].

These types of brominated porphyrins can then participate in coupling reactions such as Suzuki–Miyaura, Buchwald–Hartwig, Stille,

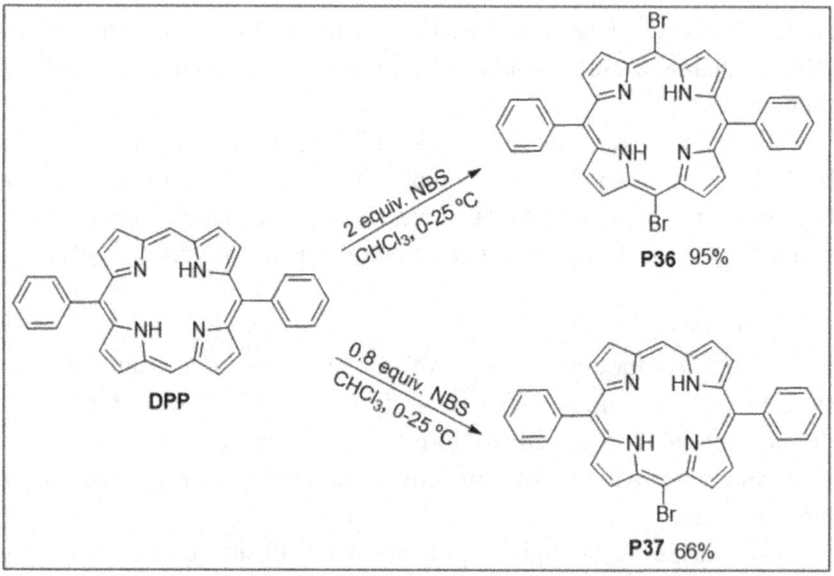

SCHEME 2.28 Examples of mono- and di-bromination of 5,15-diphenylporphyrin (**DPP**) free *meso* positions with NBS.

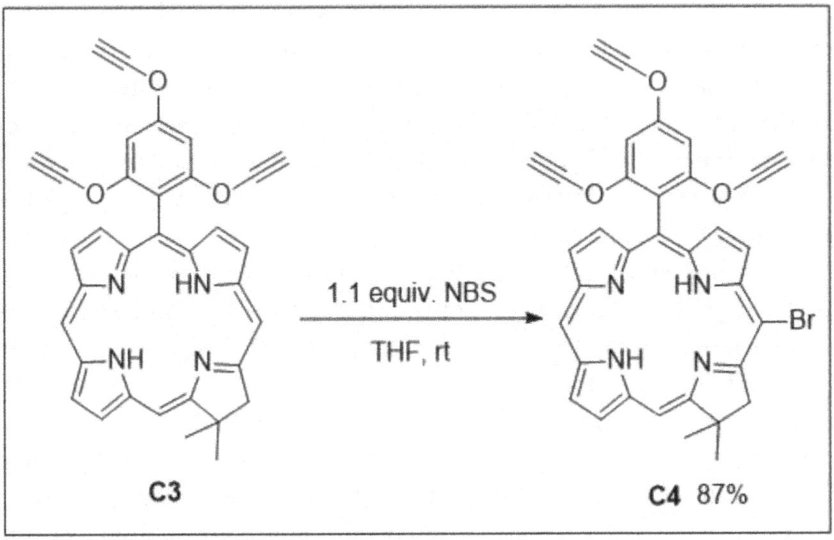

SCHEME 2.29 Mono-bromination of **C3** at its *meso*-position.

Mizoroki–Heck, and Sonogashira, among others [12,125,126]. For instance, **C3** was used in a Suzuki–Miyaura coupling for the formation of a bioconjugate, as will be described in **Chapter 4** [102].

2.5.5 Functionalization of Naturally Derived Porphyrins

Naturally derived porphyrins, in particular protoporphyrin IX, are attractive compounds for bioconjugation due to their biocompatibility and ubiquitous presence in both plants and animals, as discussed in the introductory subsection. From a chemical standpoint, protoporphyrin IX possesses carboxylic acid groups, which are ideal for the preparation of bioconjugates. Herein we will present two examples of protoporphyrin IX carboxylic acid modifications aiming at the introduction of either nucleophilic or electrophilic groups more suitable for bioconjugation.

Zhang reported a protoporphyrin IX derivative and its conjugation with PEG moieties for use as a drug for the treatment of acetaminophen-induced acute liver failure [127]. Protoporphyrin IX can be obtained through an acidic demetallation of heme, which in turn is extracted from animal blood [128]. Its modification for PEG linking encompassed the activation of the two terminal carboxylic groups by reacting them with ethyl chloroformate, resulting in the formation of the corresponding anhydrides (**P38**) (**Scheme 2.30**). This reaction follows the typical carbonyl addition-elimination reaction mechanism, where the carboxylic acid is the intervening nucleophile. The increasing anhydride reactivity towards nucleophilic attack allows the reaction with a mono-Boc protected ethylenediamine, in DMF, at room temperature, for 120 min, with formation of an amide (**P39**). Finally, the amine groups were deprotected using TFA at 0°C, selectively cleaving the Boc groups without amide hydrolysis, giving **P40** with 52% overall yield [129].

Another strategy for protoporphyrin IX modification, reported by *Xing* [130], consists in the activation of the carboxylic acid with 1-ethyl-3-(3-dimethylaminopropyl)carbodiimide (EDC) and N-hydroxysuccinimide (HOSu), followed by the addition of an amino-functionalized maleimide, with consequent formation of amide bonds (**Scheme 2.31**). Through a stoichiometric control, the di- and mono-functionalized **P41** and **P42** were obtained in 73% and 43% yields, respectively. These maleimide groups are especially reactive towards -SH-containing groups, such as cysteine

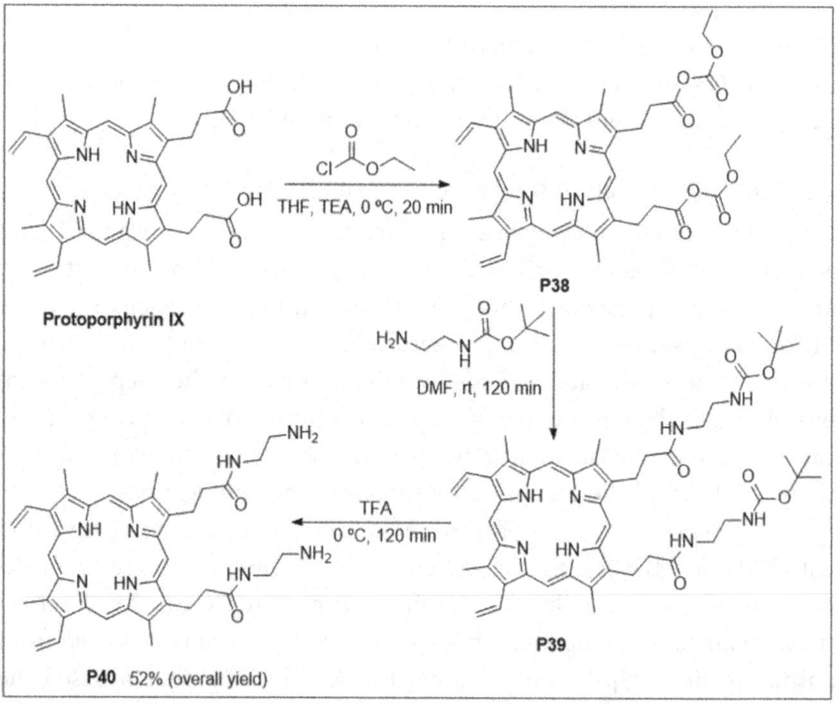

SCHEME 2.30 Synthesis of a protoporphyrin IX derivative containing two free amine groups for bioconjugation (**P40**).

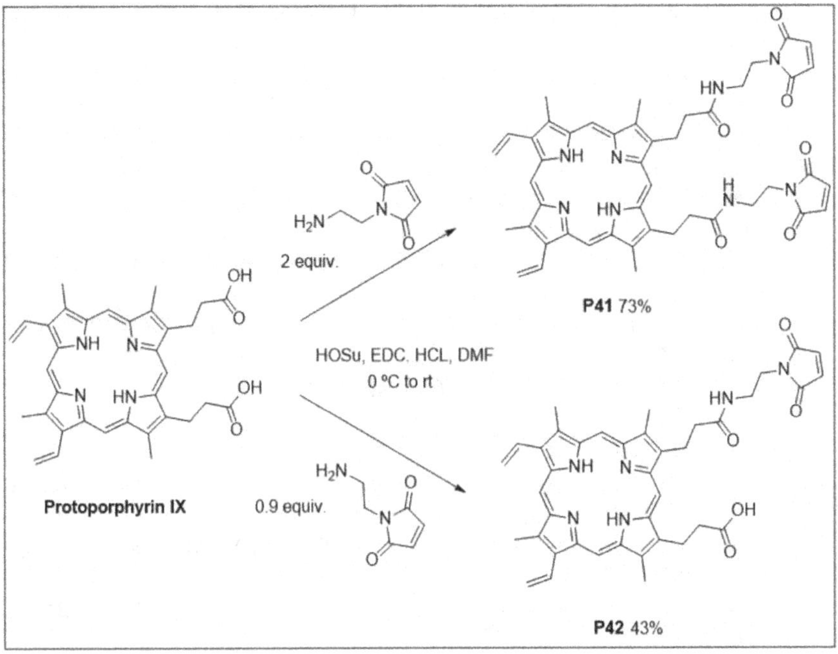

SCHEME 2.31 Functionalization of protoporphyrin IX with maleimide groups aiming a subsequent conjugation with -SH-containing biomolecules.

residues, and their conjugation with cationic peptides for use as antimicrobials will be discussed in **Chapter 6.**

NOTE

1 This chapter was written by Rafael T. Aroso and Mariette M. Pereira.

REFERENCES

[1] Fischer, H., Orth, H., *Die Chemie des Pyrrols*. Akad, **1934**, Verlagsges.
[2] Moss, G. P., *Pure Appl. Chem.*, **1987**, *59*, 779–832.
[3] Lash, T. D., *J. Porphyr. Phthalocyanines*, **2011**, *15*, 1093–1115.
[4] Battersby, A. R., *Nat. Prod. Rep.*, **2000**, *17*, 507–526.
[5] Poulos, T. L., *Chem. Rev.*, **2014**, *114*, 3919-3962.
[6] Borbas, K. E., Chapter 181: Chlorins, in *Handbook of Porphyrin Science: With Applications to Chemistry, Physics, Materials Science, Engineering, Biology and Medicine—Volume 36: BODIPYs and Chlorins: Powerful Related Porphyrin Fluorophores*, edited by Kadish, K. M., Smith, K. M., Guilard, R., World Scientific, **2016**, 1–149.
[7] Senge, M. O., Sergeeva, N. N., Hale, K. J., *Chem. Soc. Rev.*, **2021**, *50*, 4730–4789.

[8] Pereira, M. M., Monteiro, C. J. P., Peixoto, A. F., Meso-Substituted Porphyrin Synthesis from Monopyrrole: An Overview, in *Targets in Heterocyclic Systems: Chemistry and Properties*, Vol. 12, Italian Society of Chemistry, **2008**.

[9] Vicente, M., Smith, K., *Curr. Org. Synth.*, **2014**, *11*, 3–28.

[10] Pinto, S. M. A., Vinagreiro, C. S., Tomé, V. A., Piccirillo, G., Damas, L., Pereira, M. M., *J. Porphyr. Phthalocyanines*, **2019**, *23*, 329–346.

[11] Pinto, S. M. A., Henriques, C. A., Tomé, V. A., Vinagreiro, C. S., Calvete, M. J. F., Dąbrowski, J. M., Piñeiro, M., Arnaut, L. G., Pereira, M. M., *J. Porphyr. Phthalocyanines*, **2016**, *20*, 45–60.

[12] Hiroto, S., Miyake, Y., Shinokubo, H., *Chem. Rev.*, **2016**, *117*, 2910–3043.

[13] Koifman, O. I., Ageeva, T. A., *Russ. J. Org. Chem.*, **2022**, *58*, 443–479.

[14] Song, H., Liu, Q., Xie, Y., *ChemComm.*, **2018**, *54*, 1811–1824.

[15] Barona-Castaño, J., Carmona-Vargas, C., Brocksom, T., de Oliveira, K., *Molecules*, **2016**, *21*, 310.

[16] Pereira, M. M., Dias, L. D., Calvete, M. J. F., *ACS Catal.*, **2018**, *8*, 10784–10808.

[17] Calvete, M. J. F., Piñeiro, M., Dias, L. D., Pereira, M. M., *ChemCatChem*, **2018**, *10*, 3615–3635.

[18] Park, J. M., Hong, K.-I., Lee, H., Jang, W. D., *Acc. Chem. Res.*, **2021**, *54*, 2249–2260.

[19] Paolesse, R., Nardis, S., Monti, D., Stefanelli, M., Di Natale, C., *Chem. Rev.*, **2016**, *117*, 2517–2583.

[20] Kou, J., Dou, D., Yang, L., *Oncotarget*, **2017**, *8*, 81591–81603.

[21] Dabrowski, J. M., Arnaut, L. G., *Photochem. Photobiol. Sci.*, **2015**, *14*, 1765–1780.

[22] Frochot, C., Mordon, S., *J. Porphyr. Phthalocyanines*, **2019**, *23*, 347–357.

[23] Hu, X., Huang, Y.-Y., Wang, Y., Wang, X., Hamblin, M. R., *Front. Microbiol.*, **2018**, *9*, 1299.

[24] Aroso, R. T., Schaberle, F. A., Arnaut, L. G., Pereira, M. M., *Photochem. Photobiol. Sci.*, **2021**, *20*, 1497–1545.

[25] Vinagreiro, C. S., Zangirolami, A., Schaberle, F. A., Nunes, S. C. C., Blanco, K. C., Inada, N. M., da Silva, G. J., Pais, A. A. C. C., Bagnato, V. S., Arnaut, L. G., Pereira, M. M., *ACS Infect. Dis.*, **2020**, *6*, 1517–1526.

[26] Aroso, R. T., Dias, L. D., Blanco, K. C., Soares, J. M., Alves, F., da Silva, G. J., Arnaut, L. G., Bagnato, V. S., Pereira, M. M., *J. Photochem. Photobiol. B, Biol.*, **2022**, *233*, 112499.

[27] Klausen, M., Ucuncu, M., Bradley, M., *Molecules*, **2020**, *25*, 5239.

[28] Cieplik, F., Deng, D., Crielaard, W., Buchalla, W., Hellwig, E., Al-Ahmad, A., Maisch, T., *Crit. Rev. Microbiol.*, **2018**, *44*, 571–589.

[29] Liu, T. W., Huynh, E., MacDonald, T. D., Zheng, G., *Cancer Theranostics*, Elsevier Inc., **2014**, 229–254.

[30] Tsolekile, N., Nelana, S., Oluwafemi, O. S., *Molecules*, **2019**, *24*, 2669.

[31] Ethirajan, M., Chen, Y., Joshi, P., Pandey, R. K., *Chem. Soc. Rev.*, **2011**, *40*, 340–362.

[32] Geraldes, C. F. G. C., Castro, M. M. C. A., Peters, J. A., *Coord. Chem. Rev.*, **2021**, *445*, 214069.

[33] Calvete, M. J. F., Pinto, S. M. A., Pereira, M. M., Geraldes, C. F. G. C., *Coord. Chem. Rev.*, **2017**, *333*, 82–107.
[34] Pinto, S. M. A., Calvete, M. J. F., Ghica, M. E., Soler, S., Gallardo, I., Pallier, A., Laranjo, M. B., Cardoso, A. M. S., Castro, M. M. C. A., Brett, C. M. A., Pereira, M. M., Tóth, É., Geraldes, C. F. G. C., *Dalton Trans.*, **2019**, *48*, 3249–3262.
[35] Imran, M., Ramzan, M., Qureshi, A., Khan, M., Tariq, M., *Biosensors*, **2018**, *8*, 95.
[36] Senge, M. O., Brandt, J. C., *Photochem. Photobiol.*, **2011**, *87*, 1240–1296.
[37] Brandis, A., Mazor, O., Neumark, E., Rosenbach-Belkin, V., Salomon, Y., Scherz, A., *Photochem. Photobiol.*, **2005**, *81*, 983.
[38] Santos, L. L., Oliveira, J., Monteiro, E., Santos, J., Sarmento, C., *Case Rep. Oncol.*, **2018**, *11*, 769–776.
[39] Gonsalves, A. M. D. R., Pereira, M. M., *J. Heterocycl. Chem.*, **1985**, *22*, 931.
[40] Rothemund, P., *J. Am. Chem. Soc.*, **1935**, *57*, 2010–2011.
[41] Rothemund, P., *J. Am. Chem. Soc.*, **1936**, *58*, 625–627.
[42] Rothemund, P., Menotti, A. R., *J. Am. Chem. Soc.*, **1941**, *63*, 267–270.
[43] Lindsey, J. S., Schreiman, I. C., Hsu, H. C., Kearney, P. C., Marguerettaz, A. M., *J. Org. Chem.*, **1987**, *52*, 827–836.
[44] Adler, A. D., Longo, F. R., Finarelli, J. D., Goldmacher, J., Assour, J., Korsakoff, L., *J. Org. Chem.*, **1967**, *32*, 476–476.
[45] Rumyantseva, V. D., Gorshkova, A. S., Mironov, A. F., *Macroheterocycles*, **2013**, *6*, 59–61.
[46] Bhyrappa, P., Young, J. K., Moore, J. S., Suslick, K. S., *J. Am. Chem. Soc.*, **1996**, *118*, 5708–5711.
[47] van Nunen, J. L. M., Folmer, B. F. B., Nolte, R. J. M., *J. Am. Chem. Soc.*, **1997**, *119*, 283–291.
[48] Lorenz, P., Luchs, T., Hirsch, A., *Chem. Eur. J.*, **2021**, *27*, 4993–5002.
[49] Kitaoka, S., Nobuoka, K., Ihara, K., Ishikawa, Y., *RSC Adv.*, **2014**, *4*, 26777–26782.
[50] Umezawa, N., Matsumoto, N., Iwama, S., Kato, N., Higuchi, T., *Bioorg. Med. Chem.*, **2010**, *18*, 6340–6350.
[51] Odai, S., Ito, H., Kamachi, T., *J. Clin. Biochem. Nutr.*, **2019**, *65*, 178–184.
[52] Briñas, R. P., Troxler, T., Hochstrasser, R. M., Vinogradov, S. A., *J. Am. Chem. Soc.*, **2005**, *127*, 11851–11862.
[53] Asayama, S., Mizushima, K., Nagaoka, S., Kawakami, H., *Bioconju. Chem.*, **2004**, *15*, 1360–1363.
[54] Gonsalves, A. M. d'A. R., Varejão, J. M. T. B., Pereira, M. M., *J. Heterocycl. Chem.*, **1991**, *28*, 635–640.
[55] Tsuchida, E., Hasegawa, E., Kanayama, T., *Macromolecules*, **1978**, *11*, 947–955.
[56] Stäubli, B., Fretz, H., Piantini, U., Woggon, W.-D., *Helv. Chim. Acta*, **1987**, *70*, 1173–1193.
[57] Kim, J. B., Leonard, J. J., Longo, F. R., *J. Am. Chem. Soc.*, **1972**, *94*, 3986–3992.
[58] Hill, C. L., Williamson, M. M., *J. Chem. Soc., Chem. Commun.*, **1985**, 1228–1229.

[59] Silva, M., Fernandes, A., Bebiano, S. S., Calvete, M. J. F., Ribeiro, M. F., Burrows, H. D., Pereira, M. M., *ChemComm.*, **2014**, *50*, 6571–6573.
[60] Henriques, C. A., Pinto, S. M. A., Aquino, G. L. B., Pineiro, M., Calvete, M. J. F., Pereira, M. M., *ChemSusChem*, **2014**, *7*, 2821–2824.
[61] Fareghi-Alamdari, R., Golestanzadeh, M., Bagheri, O., *RSC Adv.*, **2016**, *6*, 108755–108767.
[62] Grancho, J. C. P., Pereira, M. M., Miguel, M. da G., Gonsalves, A. M. R., Burrows, H. D., *Photochem. Photobiol.*, **2007**, *75*, 249–256.
[63] Gao, Y., Pan, J. G., Huang, Y. J., Ding, S. Y., Wang, M. L., *J. Porphyr. Phthalocyanines*, **2015**, *19*, 1251–1255.
[64] Johnstone, A. W. R., Luisa, P. G., Nunes, M., Pereira, M., Rocha Gonsalves, M. d' A., Serra, C. A., *Heterocycles*, **1996**, *43*, 1423.
[65] Golf, H. R. A., Reissig, H.-U., Wiehe, A., *J. Org. Chem.*, **2015**, 1548–1568.
[66] Rodrigues, J. M. M., Farinha, A. S. F., Muteto, P. V., Woranovicz-Barreira, S. M., Almeida Paz, F. A., Neves, M. G. P. M. S., Cavaleiro, J. A. S., Tomé, A. C., Gomes, M. T. S. R., Sessler, J. L., Tomé, J. P. C., *ChemComm*, **2014**, *50*, 1359–1361.
[67] Vinagreiro, C. S., Gonçalves, N. P. F., Calvete, M. J. F., Schaberle, F. A., Arnaut, L. G., Pereira, M. M., *J. Fluor. Chem.*, **2015**, *180*, 161–167.
[68] Hirohara, S., Oka, C., Totani, M., Obata, M., Yuasa, J., Ito, H., Tamura, M., Matsui, H., Kakiuchi, K., Kawai, T., Kawaichi, M., Tanihara, M., *J. Med. Chem.*, **2015**, *58*, 8658–8670.
[69] Avner, Y., David, A., Zeev, G., *Pharmaceutical Compositions Comprising Porphyrins and Some Novel Porphyrin Derivatives*, Technion Research and Development Foundation Ltd, Yeda Research and Development Co Ltd, **2004**, US6730666B1.
[70] Silva, A. M. G., Tomé A. C., Neves, M. G. P. M. S., Silva, A. M. S., Cavaleiro, J. A. S., *J. Org. Chem.*, **2005**, *70*, 2306–2314.
[71] Lindsey, J. S., Hsu, H. C., Schreiman, I. C., *Tetrahedron Lett.*, **1986**, *27*, 4969–4970.
[72] Wagner, R. W., Lawrence, D. S., Lindsey, J. S., *Tetrahedron Lett.*, **1987**, *28*, 3069–3070.
[73] Lindsey, J. S., Schreiman, I. C., Hsu, H. C., Kearney, P. C., Marguerettaz, A. M., *J. Org. Chem.*, **1987**, *52*, 827–836.
[74] Lindsey, J. S., Wagner, R. W., *J. Org. Chem.*, **1989**, *54*, 828–836.
[75] Pollak, K. W., Sanford, E. M., Fréchet, J. M. J., *J. Mater. Chem.*, **1998**, *8*, 519–527.
[76] Soman, R., Raghav, D., Sujatha, S., Rathinasamy, K., Arunkumar, C., *RSC Adv.*, **2015**, *5*, 61103–61117.
[77] Peters, M. K., Röhricht, F., Näther, C., Herges, R., *Org. Lett.*, **2018**, *20*, 7879–7883.
[78] Rosa, M., Jędryka, N., Skorupska, S., Grabowska-Jadach, I., Malinowski, M., *J. Mol. Sci.*, **2022**, *23*, 11321.
[79] Plunkett, S., Senge, M. O., *ECS Tran.*, **2011**, *35*, 147–157.
[80] Lindsey, J. S., Prathapan, S., Johnson, T. E., Wagner, R. W., *Tetrahedron*, **1994**, *50*, 8941–8968.

[81] Shieh, M.-J., Peng, C.-L., Lou, P.-J., Chiu, C.-H., Tsai, T.-Y., Hsu, C.-Y., Yeh, C.-Y., Lai, P.-S., *J. Control. Release*, **2008**, *129*, 200–206.
[82] Fagadar-Cosma, E., Vlascici, D., Birdeanu, M., Fagadar-Cosma, G., *Arab. J. Chem.*, **2019**, *12*, 1587–1594.
[83] Slomp, A. M., Barreira, S. M. W., Carrenho, L. Z. B., Vandresen, C. C., Zattoni, I. F., Ló, S. M. S., Dallagnol, J. C. C., Ducatti, D. R. B., Orsato, A., Duarte, M. E. R., Noseda, M. D., Otuki, M. F., Gonçalves, A. G., *Bioorg. Med. Chem.*, **2017**, *27*, 156–161.
[84] Séverac, M., Pleux, L. L., Scarpaci, A., Blart, E., Odobel, F., *Tetrahedron Lett.*, **2007**, *48*, 6518–6522.
[85] Feese, E., Gracz, H. S., Boyle, P. D., Ghiladi, R. A., *J. Porphyr. Phthalocyanines*, **2019**, *23*, 1414–1439.
[86] Lindsey, J. S., *Acc. Chem. Res.*, **2009**, *43*, 300–311.
[87] Ryppa, C., Senge, M. O., Hatscher, S. S., Kleinpeter, E., Wacker, P., Schilde, U., Wiehe, A., *Chem. Eur. J.*, **2005**, *11*, 3427–3442.
[88] Senge, M. O., *Acc. Chem. Res.*, **2005**, *38*, 733–743.
[89] Vicente, M., Smith, K., *Curr. Org. Synth.*, **2014**, *11*, 3–28.
[90] Laha, J. K., Dhanalekshmi, S., Taniguchi, M., Ambroise, A., Lindsey, J. S., *Org. Process. Res. Dev.*, **2003**, *7*, 799–812.
[91] Pereira, N. A. M., Pinho e Melo, T. M. V. D., *Org. Prep. Proced. Int.*, **2014**, *46*, 183–213.
[92] Thamyongkit, P., Bhise, A. D., Taniguchi, M., Lindsey, J. S., *J. Org. Chem.*, **2006**, *71*, 903–910.
[93] Lee, C.-H., Lindsey, S. J., *Tetrahedron*, **1994**, *50*, 11427–11440.
[94] Tamaru, S., Yu, L., Youngblood, W. J., Muthukumaran, K., Taniguchi, M., Lindsey, J. S., *J. Org. Chem.*, **2004**, *69*, 765–777.
[95] Rao, P. D., Littler, B. J., Geier, G. R., Lindsey, J. S., *J. Org. Chem.*, **2000**, *65*, 1084–1092.
[96] Geier III, G. R., Callinan, J. B., Dharma Rao, P., Lindsey, J. S., *J. Porphyr. Phthalocyanines*, **2001**, *5*, 810–823.
[97] Dogutan, D. K., Ptaszek, M., Lindsey, J. S., *J. Org. Chem.*, **2008**, *73*, 6187–6201.
[98] Lee, C.-H., Li, F., Iwamoto, K., Dadok, J., Bothner-By, A. A., Lindsey, J. S., *Tetrahedron*, **1995**, *51*, 11645–11672.
[99] Rao, P. D., Dhanalekshmi, S., Littler, B. J., Lindsey, J. S., *J. Org. Chem.*, **2000**, *65*, 7323–7344.
[100] Lin, T., Shang, X. S., Adisoejoso, J., Liu, P. N., Lin, N., *J. Am. Chem. Soc.*, **2013**, *135*, 3576–3582.
[101] She, Y. B., Fu, H. Y., Zhang, Z. L., Song, X.-F., Sun, Z.-C., *Heterocycles*, **2014**, *89*, 503.
[102] Matsumoto, N., Taniguchi, M., Lindsey, J. S., *J. Porphyr. Phthalocyanines*, **2020**, *24*, 362–378.
[103] Liu, M., Chen, C.-Y., Mandal, A. K., Chandrashaker, V., Evans-Storms, R. B., Pitner, J. B., Bocian, D. F., Holten, D., Lindsey, J. S., *New J. Chem.*, **2016**, *40*, 7721–7740.

[104] Ptaszek, M., Bhaumik, J., Kim, H.-J., Taniguchi, M., Lindsey, J. S., *Org. Process Res. Dev.*, **2005**, *9*, 651–659.
[105] Taniguchi, M., Ra, D., Mo, G., Balasubramanian, T., Lindsey, J. S., *J. Org. Chem*, **2001**, *66*, 7342–7354.
[106] Mikus, A., Łopuszyńska, B., *Chem. Asian J.*, **2020**, *16*, 261–276.
[107] Catalano, M. M., Crossley, M. J., Harding, M. M., King, L. G., *J. Chem. Soc., Chem. Commun.*, **1984**, 1535–1536.
[108] Hombrecher, H. K., Gherdan, V. M., Ohm, S., Cavaleiro, J. A. S., Graça, M. da, Neves, P. M. S., Fátima Condesso, M., *Tetrahedron*, **1993**, *49*, 8569–8578.
[109] Wang, Z., Menke Nitration, in *Comprehensive Organic Name Reactions and Reagents*, edited by Wang, Z., Wiley & Sons, **2010**.
[110] Luguya, R., Jaquinod, L., Fronczek, F. R., Vicente, M. G. H., Smith, K. M., *Tetrahedron*, **2004**, *60*, 2757–2763.
[111] Sibrian-Vazquez, M., Jensen, T. J., Vicente, M. G. H., *J. Photochem. Photobiol. B*, **2007**, *86*, 9–21.
[112] Chinnusamy, T., Rodionov, V., Kühn, F. E., Reiser, O., *Adv. Synth. Catal.*, **2012**, *354*, 1827–1831.
[113] Joshi, S. M., de Cózar, A., Gómez-Vallejo, V., Koziorowski, J., Llop, J., Cossío, F. P., *ChemComm*, **2015**, *51*, 8954–8957.
[114] Busby, C. A., Dinello, R. K., Dolphin, D., *Can. J. Chem.*, **1975**, *53*, 1554–1555.
[115] Morozov, V. V., Semeikin, A. S., Gnedin, B. G., Berezin, B. D., *Chem. Heterocycl. Compounds*, **1988**, *24*, 628–630.
[116] Gonsalves, A. M. D. R., Johnstone, R. A. W., Pereira, M. M., SantAna, A. M. P., Serra, A. C., Sobral, A. J. F. N., Stocks, P. A., *Heterocycles*, **1996**, *43*, 829.
[117] Clayden, J., Greeves, N., Warren, S., *Organic Chemistry*, 2nd Edition, Oxford University Press, **2012**, 485–486.
[118] Monteiro, C. J. P., Pereira, M. M., Pinto, S. M. A., Simões, A. V. C., Sá, G. F. F., Arnaut, L. G., Formosinho, S. J., Simões, S., Wyatt, M. F., *Tetrahedron*, **2008**, *64*, 5132–5138.
[119] Pereira, M. M., Monteiro, C. J. P., Simões, A. V. C., Pinto, S. M. A., Abreu, A. R., Sá, G. F. F., Silva, E. F. F., Rocha, L. B., Dąbrowski, J. M., Formosinho, S. J., Simões, S., Arnaut, L. G., *Tetrahedron*, **2010**, *66*, 9545–9551.
[120] Simões, A. V. C., Adamowicz, A., Dąbrowski, J. M., Calvete, M. J. F., Abreu, A. R., Stochel, G., Arnaut, L. G., Pereira, M. M., *Tetrahedron*, **2012**, *68*, 8767–8772.
[121] Locos, O. B., Arnold, D. P., *Org. Biomol. Chem.*, **2006**, *4*, 902.
[122] Odobel, F., Suresh, S., Blart, E., Nicolas, Y., Quintard, J.-P., Janvier, P., Le Questel, J.-Y., Illien, B., Rondeau, D., Richomme, P., Häupl, T., Wallin, S., Hammarström, L., *Chem. Eur. J.*, **2002**, *8*, 3027.
[123] Kato, A., Hartnell, R. D., Yamashita, M., Miyasaka, H., Sugiura, K., Arnold, D. P., *J. Porphyr. Phthalocyanines*, **2004**, *8*, 1222–1227.
[124] Chumakov, D. E., Khoroshutin, A. V., Anisimov, A. V., Kobrakov, K. I., *Chem. Heterocycl. Compounds*, **2009**, *45*, 259–283.

[125] Wang, K., Osuka, A., Song, J., *ACS Cent. Sci.*, **2020**, *6*, 2159–2178.
[126] Prakash, K., Osterloh, W. R., Rathi, P., Kadish, K. M., Sankar, M., *J. Organomet. Chem.*, **2021**, *956*, 122114.
[127] Zhang, T., Fang, J., Tsutsuki, H., Ono, K., Islam, W., Sawa, T., *Biol. Pharm. Bull.*, **2019**, *42*, 1199.
[128] Smith, K. M., Porphyrins, Corrins and Phthalocyanines, in *Comprehensive Heterocyclic Chemistry*, LSU Scholarly Repository, **1984**, 377–442.
[129] Sahoo, S. K., Sawa, T., Fang, J., Tanaka, S., Miyamoto, Y., Akaike, T., Maeda, H., *Bioconjug. Chem.*, **2002**, *13*, 1031–1038.
[130] Liu, F., Ni, A. S. Y., Lim, Y., Mohanram, H., Bhattacharjya, S., Xing, B., *Bioconjug. Chem.*, **2012**, *23*, 1639.

CHAPTER 3

Porphyrin-Based Dendrimer Bioconjugates

3.1 INTRODUCTION

Branching is one of the most frequent growth approaches in nature. We can observe this pattern in trees, river networks, neurons, lungs, and kidneys, for instance. Branched dendritic structures grow in every sphere of nature, where the multiplicity of terminal functionalities may lead to enhanced properties and functions (**Figure 3.1**).

The 1970s are considered the birth of branching in chemistry, since new classes of highly branched synthetic macromolecules have found their first applications during this period. In this regard, *Vogtle* [4] developed the first systematic synthetic methodology to obtain branched amines with low molecular weight (M_W), which were designated by "cascade" molecules. Later, in 1985, inspired by natural structures, *Tomalia* [5,6] described an efficient synthetic methodology to obtain macromolecules using a repetitive growth pathway. These macromolecules were chemicals denominated by *dendrimers*, derived from the Greek word *dendron*, which means "tree." Nowadays, the term *dendrimer* is attributed to a well-defined multi-functionalized molecule, which encompasses nanometer branches. *Dendrimers*' molecular structures have a central core or nucleus and branches with terminal functional groups on the outer surface (**Figure 3.2**).

DOI: 10.1201/9781003119265-3

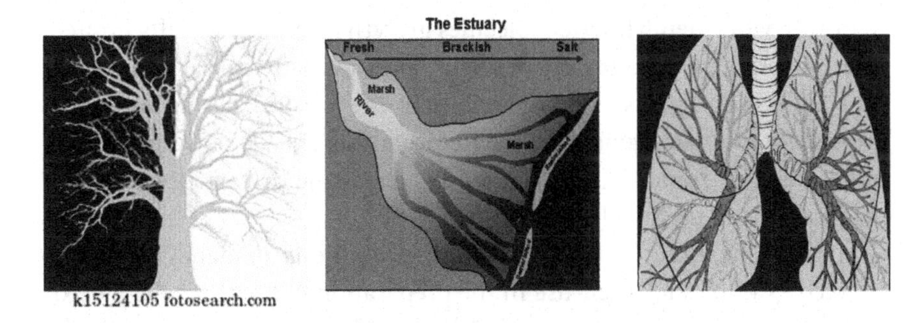

FIGURE 3.1 Examples of dendritic structures in nature: (a) a tree; (b) a river estuary; (c) blood vessel ramifications in the lung.

Source: Adapted from [1,2,3].

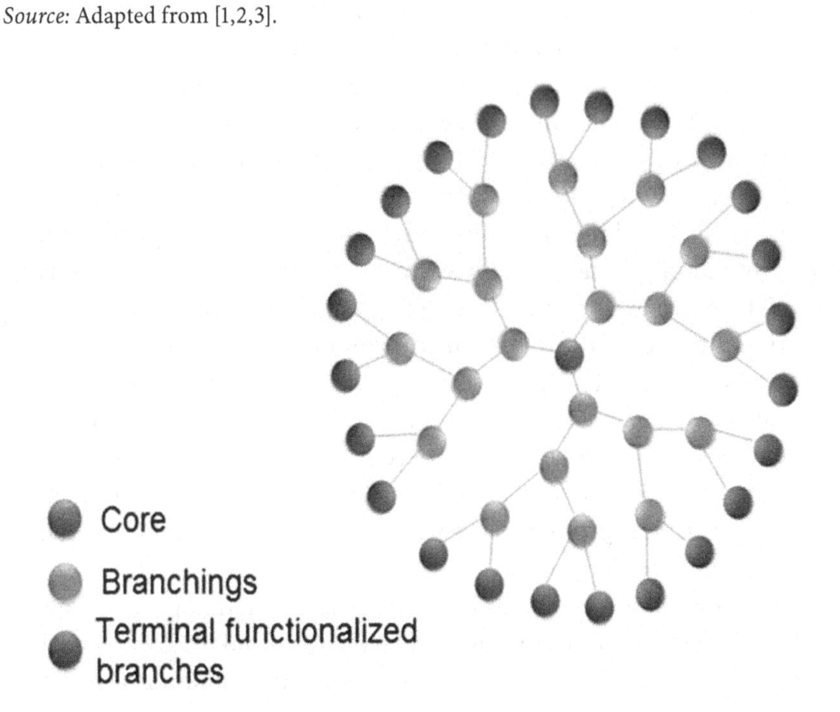

FIGURE 3.2 Dendrimer structure.

Typically, dendrimers with well-defined sizes and structures are obtained using a stepwise chemical synthetic approach, and contrary to other polymers, they possess a low polydispersity index. The molecular weight of linear hyperbranched polymers is difficult to control (infinite in limit!), while dendrimeric molecules growth is mathematically limited. Interest for these structurally defined dendrimeric terminal functionalized molecules increased exponentially in the last decades as a result of their several applications in fields like catalysis [7,8,9], sensing [7,10,11], molecular electronics [7,12,13], drug delivery, gene delivery [14,15,16,17], diagnostic, and therapy [14,18,19,20].

Herein, the main approaches for the synthesis of dendrimers and selected examples for their use in the preparation of tetrapyrrolic macrocycles bioconjugates are described. Additionally, a brief discussion of their applications in medicinal chemistry and issues regarding biodistribution and organism elimination is also presented.

There are two main synthetic strategies for dendrimer preparation, based on divergent and convergent synthesis [21]. Both have their intrinsic advantages and limitations, and before designing a synthetic methodology for obtaining a dendrimer, its structure and application must be considered [22]. The divergent growth methodology was firstly developed by *Tomalia* in the 1980s [5,6]. It starts by reacting the peripherical functional groups of the core with the adequate reactive group of the monomer, giving structures considered of first generation (**G1**). After performing deprotection/activation of the peripherical functionalities of **G1**, a reaction to promote the formation of a new covalent linkage with the monomer generates dendrimers of second generation (**G2**), and so on (**Figure 3.3**) [23]. In sum, this synthetic strategy basically involves two different steps: (a) covalent linking of the monomer with the core; (b) deprotection/activation of monomer-ending functional groups to create a functionality capable to couple with a new monomer [23,24].

In this methodology, it is important that each reaction proceeds quantitatively at each coupling and activation step to avoid deficiently formed branches [23,25]. This is progressively difficult for higher-generation dendrimers, since there is an exponential increase in the number of terminal functional groups, making these higher-generation dendrimers quite dense and with high steric hindrance, which consequently lessens the overall yield. In addition, since structurally perfect dendrimers and defective dendrimers are similar and with approximately equal size, the application of standard purification and separation strategies may not be efficient. It should be emphasized that, despite its intrinsic drawbacks, the divergent growth strategy presents several positive points

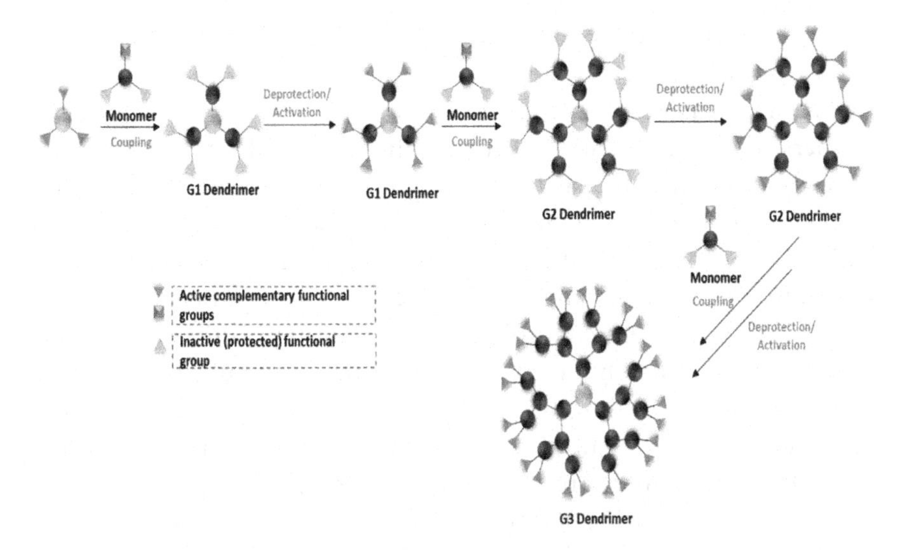

FIGURE 3.3 Dendrimer divergent growth synthetic strategy.

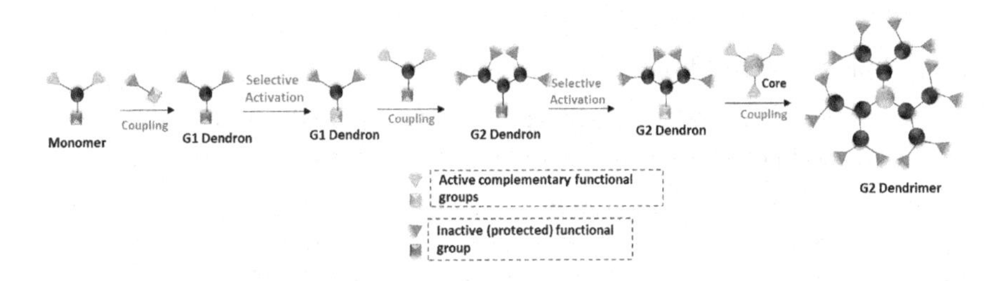

FIGURE 3.4 Dendrimer convergent growth synthetic strategy.

that turn it into the main synthetic *via* obtaining commercially available dendrimers [26].

The convergent dendrimer growth strategy (**Figure 3.4**) was designed and developed by *Fréchet* in the 1990s, and in this case, the growth occurs from the surface to the core of the molecule [27,28,29]. Under these conditions, perfectly branched dendrons (dendrimer wedges) are obtained through coupling and activation reactions until the intended size is attained. After this iterative process, coupling of the dendron to the polyfunctional core is performed, giving the desired dendrimer

[30]. Since this strategy involves fewer reactions during the coupling and activation steps, a higher structural control and dendrimers with higher monodispersity are obtained. In addition, asymmetric dendrimers, obtained by attachment of different dendrons to the polyfunctional nucleus, are easily purified and characterized. This may be attributed to the use of equimolar quantities/slight excess of reactants which input considerable differences between polarity and molar masses of dendron/dendrimers and side products. The main drawback of this methodology that limits its commercial application is the decrease on the reactivity/availability of the functionalities and bulkier dendrons, which can cause shielding of the polyfunctional core, giving incomplete substitutions. In general, with the convergent growth strategy, lower-generation dendrimers are obtained when comparing with the divergent growth strategy [26,30,31].

Although highly implemented, both strategies have similar disadvantages, like the need for several iterative reaction steps/purification involving a high consumption of time and a large waste of chemical and valuable starting materials.

3.1.1 Polyester Dendrimer

Since *Fréchet*'s [32] first report of a polyester dendrimer in 1992, these dendrimers have been extensively studied and mainly explored for biomedical applications [16,18,33,34]. This class of compounds presents several attractive features that make them appealable, namely, their high biocompatibility [35,36], low toxicity [37,38], and biodegradability [33,34].

Hult [39] published the synthesis of a group of aliphatic polyester dendrimers using 2,2-*bis*(hydroxymethyl)propionic acid (*bis*-MPA) as building block and 1,1,1-tris(hydroxyphenyl)ethane as core molecule. Initially, the different generation dendrons were synthesized using a route that involved the conversion of *bis*-MPA into the corresponding acetate esters (**Scheme 3.1**). Two *routes* were used: (a) *Bis*-MPA reacted with acetyl chloride in the presence of Et_3N and 4-dimethylaminopyridine (DMAP) in CH_2Cl_2 (**Scheme 3.1a**) (this reaction is based on an addition-elimination mechanism that starts with the nucleophilic attack by one of the lone pairs of the oxygen atoms of *bis*-MPA [**A**] on the electrophilic carbon atom of acetyl chloride [**B**], followed by elimination of HCl), and b) *via* S_N2 reaction of *bis*-MPA with benzyl bromide in the presence of KOH and *N,N*-dimethylformamide (DMF) as solvent, giving the corresponding benzyl ester (**Scheme 3.1b**). It should be emphasized that the preparation of the

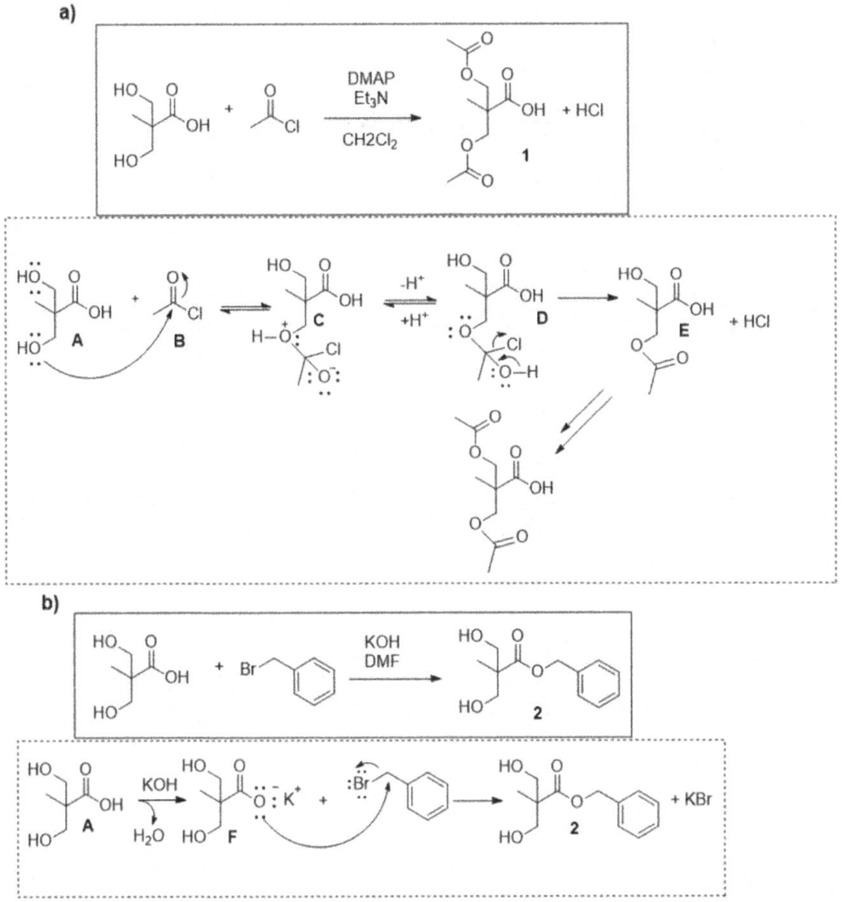

SCHEME 3.1 Synthetic *routes* and mechanisms for the conversion of *bis*-MPA into the corresponding (a) acetate esters or (b) benzyl esters.

potassium salt (**F**) preceding the addition of the benzyl bromide is crucial in order to avoid secondary reactions with *bis*-MPA hydroxyl groups.

The following step was the conversion of acetate ester **1** into its acyl chloride (**Scheme 3.2a**) using an excess of oxalyl chloride and a catalytic amount of DMF in CH_2Cl_2. In this case, the reaction mechanism involves the formation of a Vilsmeier reagent (imidoyl chloride **E**) by reaction of DMF (**A**) with oxalyl chloride (**B**). Imidoyl chloride (**E**) works as chlorinating agent that reacts with the carboxylic group to give the intended acyl chloride **3**. The coupling reaction between **2** and **3** was then performed *via* nucleophilic addition-elimination mechanism, giving the

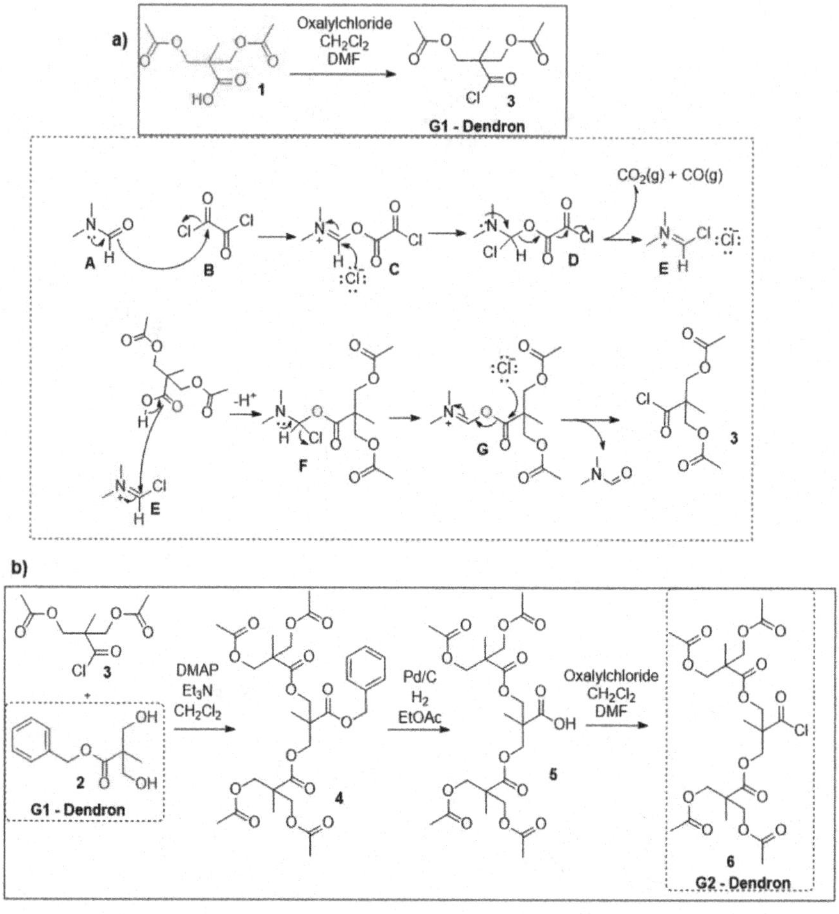

SCHEME 3.2 Synthesis and mechanisms for **G1** and **G2** generation dendrons: (a) conversion of acetate ester **1** into the acyl chloride; (b) nucleophilic addition-elimination.

benzyl ester intermediate **4**, whose deprotection *via* hydrogenolysis, using Pd/C (10%), at atmospheric pressure, afforded the carboxylic acid derivative **5 (Scheme 3.2b)**.

The last step for the dendrimer synthesis using the convergent approach was the coupling to the selected core. As an example, the synthetic procedures for the preparation of **G1** and **G2** dendrimers used by *Hult* are presented in **Scheme 3.3** [39]. In both cases, coupling of the core 1,1,1-tris(hydroxyphenyl)ethane with **G1** and **G2** dendrons is performed according to a general esterification procedure, previously described (see

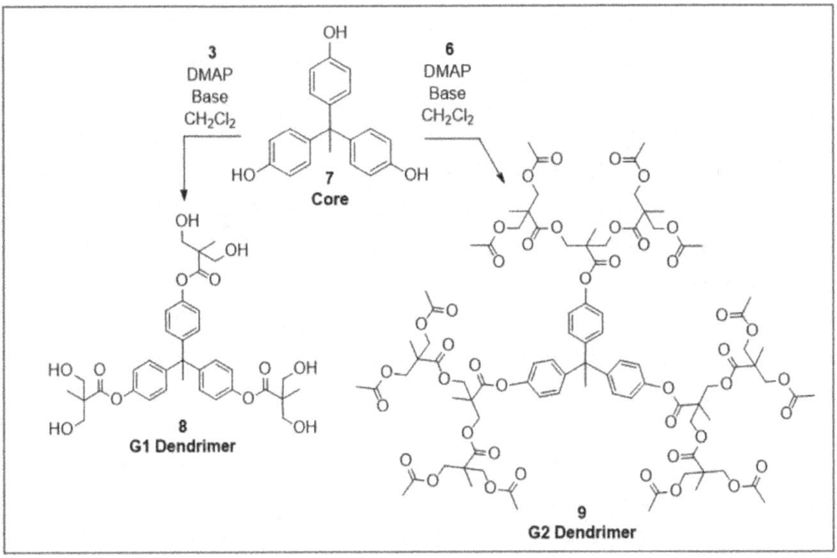

SCHEME 3.3 Synthetic *via* for **G1** and **G2** generation dendrimers.

mechanism in **Scheme 3.1**). With this strategy, monodisperse polyester dendrimers **G1** to **G4**, with 93, 89, 74, and 30% yields, with molecular weight (M_w) ranging from 906 to 7.549 g/mol, were obtained.

In 2001, *Fréchet* [40] reported the synthesis of a polyester dendrimer following a divergent growth strategy. In this synthetic route, the divergent growth begins with the synthesis of benzylidene-protected anhydride derivative **11** (**Scheme 3.4**). For this purpose, the diol group of 2,2-bis(hydroxymethyl)propionic acid was protected, using benzaldehyde dimethyl acetal and *p*-toluenesulfonic acid (TsOH) in acetone, *via* a trans-acetylation mechanism (**Scheme 3.4a**). The reaction begins with the protonation of the benzaldehyde dimethyl acetal (**A**) with *p*-toluenesulfonic acid, followed by the elimination of methanol. Then, nucleophilic addition of the alcohol yields the protected derivative **10** (**Scheme 3.4a**), which undergoes self-condensation in CH_2Cl_2, using dicyclohexylcarbodiimide (DCC) as dehydrating agent, giving the anhydride **11** (**Scheme 3.4b**). The first step of the mechanism is the deprotonation of the carboxylic group of **10**, giving intermediate **G**, followed by nucleophilic attack of the carboxylate oxygen to the electrophilic carbon of DCC, giving the intermediate *O*-acylisourea (**I**). Then, nucleophilic attack by the carboxylic group of a second molecule of **10**, followed by the liberation of one molecule of dicyclohexylurea (**L**), gives **11** (**Scheme 3.4b**).

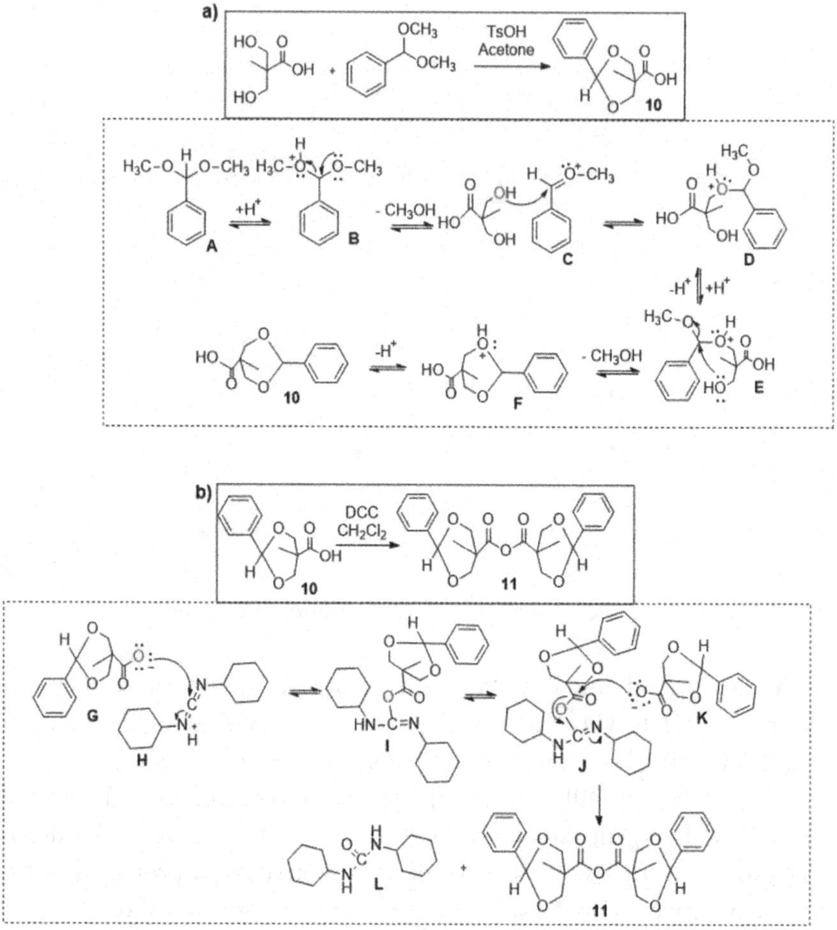

SCHEME 3.4 Synthetic route and mechanisms for **G1** and **G2** polyester dendrimers using divergent growth strategy.

Coupling with 1,1,1-*tris*-(hydroxyphenyl)ethane, using 1:3 mixture of pyridine and CH_2Cl_2 and DMAP as acylation catalyst, is performed to obtain **12**, as depicted in the mechanism presented in **Scheme 3.5**. Then, deprotection of **12** under hydrogenolysis conditions affords **G1-dendrimer 13 (Scheme 3.5)**. For further growing to obtain higher-size dendrimers, the previously described coupling/deprotection sequence is repeated.

3.1.2 Polyamidoamine (PAMAM) Dendrimers

Polyamidoamine (PAMAM) dendrimers have attracted much interest, particularly in materials and medicine areas [41,42,43,44]. Regarding

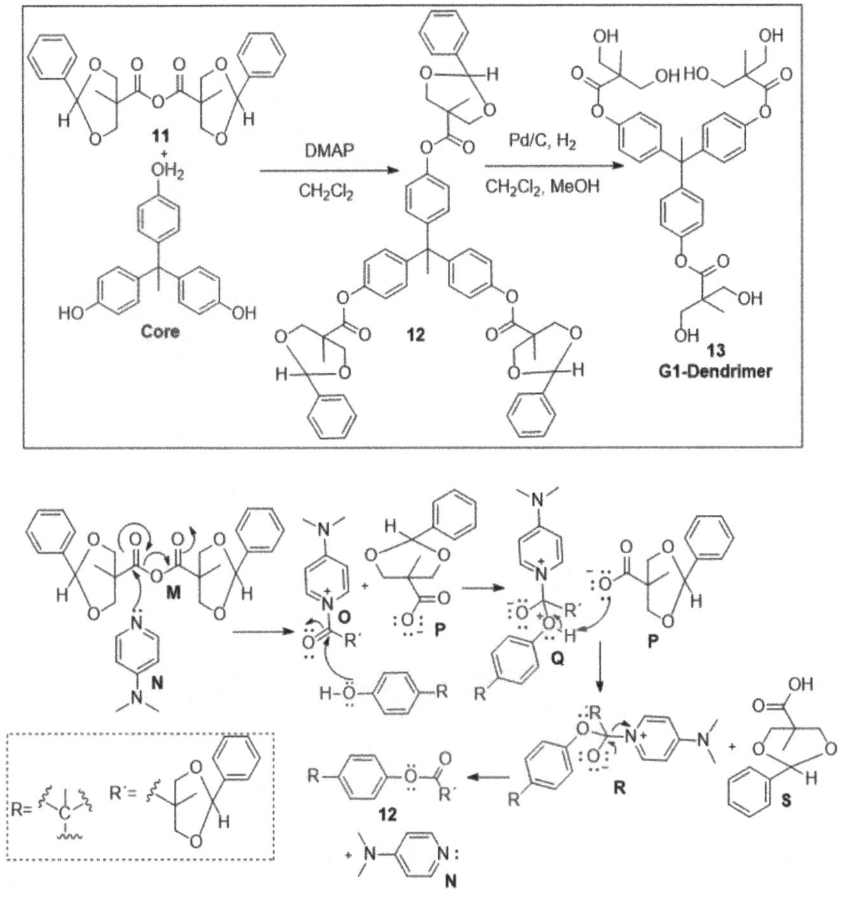

SCHEME 3.5 Synthetic route and mechanisms for **G1** and **G2** polyester dendrimers using divergent growth strategy.

medicinal applications, we highlight their use in drug delivery, molecular encapsulation, and gene therapy [42,45,46,47,48].

In 1985, *Tomalia* [5] described the first example of a PAMAM dendrimer, and since then, a growing interest related with this family has been observed [25,42,44,49,50,51,52,53]. The methodology [5,6] used to prepare this type of dendrimers is based on a divergent growth with sequential two-step reactions, under mild conditions: (a) Michael addition of ethylenediamine (central core) to methyl acrylate (**14**) gives half-generation dendrimers with terminal ester groups (**15, Scheme 3.6a**); (b) amidation of half-generation dendrimers with ethylenediamine gives **16**, which suffers a second Michael addition using methyl acrylate to give

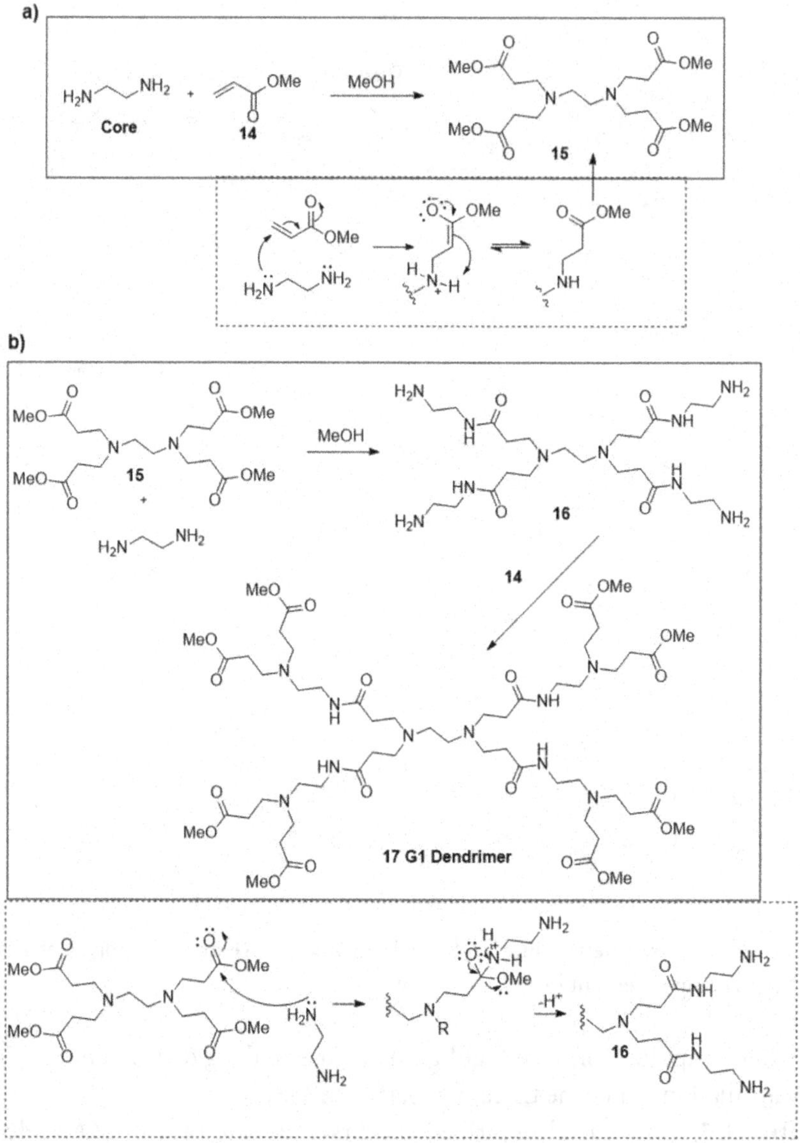

SCHEME 3.6 Aliphatic PAMAM dendrimer obtained *via* divergent growth strategy and corresponding mechanisms: (a) synthesis of half-generation dendrimer **15**; (b) synthesis of G1-dendrimer **17**.

G1 dendrimer (Scheme 3.6b). In the first step, there is the nucleophilic attack of the amine (Michael donor) to the double bond of methyl acrylate (Michael acceptor) to give the ester derivative **15 (Scheme 3.6a)**. The second step begins with the nucleophilic attack of the amine to the carbonyl group, and after the elimination of the methoxy leaving group, the corresponding amide (**16**) is formed (**Scheme 3.6b**). Dendrimer **17** (**G1**) is obtained after reaction of **16** with methyl acrylate.

This strategy allows preparation of PAMAM dendrimers in high yields, and further generations (up to 7) can be easily achieved by repeating the synthesis/purification steps. It should be mentioned that to avoid the formation of structural defects, the use of large excess of the Michael acceptor, diamine, or growth linker is a crucial issue to prepare higher-generation dendrimers [54].

PAMAM derivatives can also be obtained *via* a convergent growth [49,50,51,52]. In 2006, *Lee* [51] proposed a convergent growth strategy that starts with the synthesis of different generation dendrons using propargylamine. In this case, the growth of the dendron is achieved by a sequential synthetic strategy that involves a Michael addition, followed by an amidation, until **G1 dendron 20** is obtained (**Scheme 3.7**).

Than **G1 dendrimer 21** was prepared by coupling **G1 dendron 20** to the core, via click chemistry approach (Sharpless, Bertozzi, and Meldal were awarded with the 2022 Nobel Prize in Chemistry due to their contribution to click chemistry) [55,56,57,58] (**Scheme 3.8**). The mechanism involved in this step starts with the generation of copper acetylide intermediate (**A**) that reacts with the azide derivative, giving complex (**B**). Then (**B**) undergoes rearrangement to form a more stable intermediate (**C**). Further,

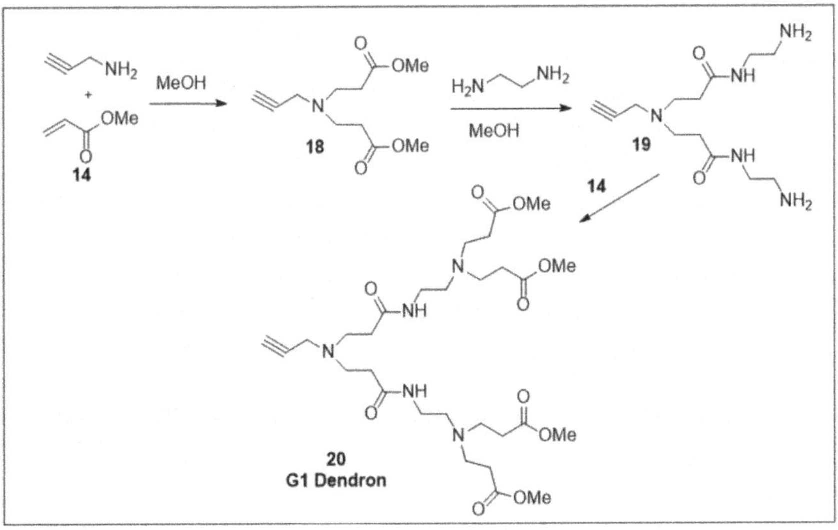

SCHEME 3.7 Aliphatic PAMAM **G1** dendron, obtained *via* convergent growth strategy.

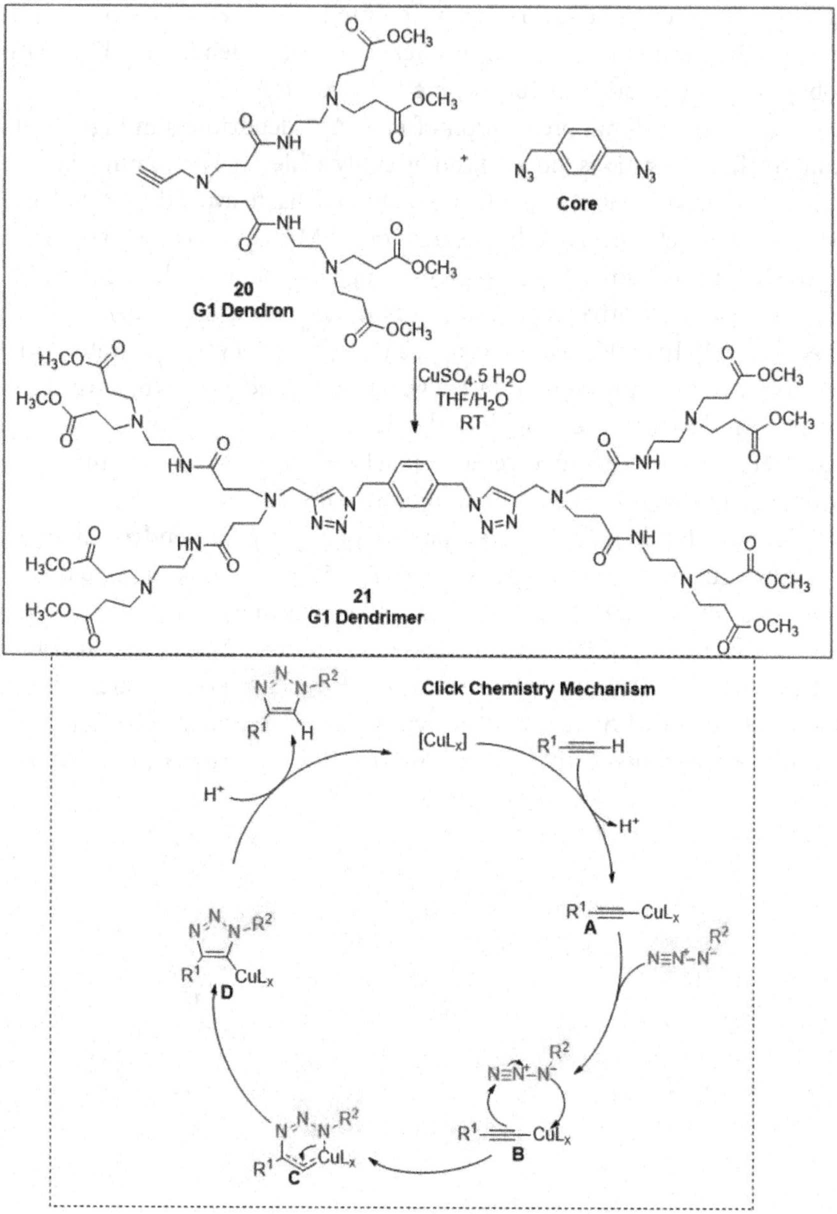

SCHEME 3.8 Aliphatic PAMAM **G1** dendrimer, obtained *via* convergent growth strategy.

intermediate (**D**) is obtained through ring contraction of (**C**). In the last step, protonation of (**D**) occurs, giving the 1,4-disubstituted 1,2,3-triazole derivative, and Cu(I) is regenerated. The authors concluded that the best conditions for coupling of **G1 dendron** with the *bis*(azide) core were the use of 2.1 equivalent of dendron, 10 mol % $CuSO_4.5H_2O$, 20 mol % sodium ascorbate in a 4:1 solvent mixture of THF to H_2O at room temperature for 1.5 h. Under these conditions, **G1 dendrimer** was obtained in up to 97% yield. They were also able to obtain higher-generation dendrimers with excellent yields (**G2:** 95%, reaction time = 2 h; **G3:** 94%, reaction time = 3 h; **G4:** 86%, reaction time = 4 h).

3.1.3 Poly(Peptide) Dendrimers (PPD)

In the early 1980s, *Denkewalter* [59,60,61,62] described the first examples of poly(peptide) dendrimers (PPD). Since then, there has been an exponential growth on reports related to this type of dendrimers [63]. Like other dendrimer families, PPDs have been studied for several medicinal applications. Particularly, drug/gene delivery and vaccine technology seem to be the driving force for the development of new PPD dendrimers [63,64,65,66,67,68,69,70].

Poly(peptide) dendrimers can be divided in two main classes according to the type of interaction between the amino acids and the dendrimer structure: (a) covalent peptide dendrimers (type I and II), when natural or non-natural amino acids are part of the framework, either in the core, branching units, or grafted onto the surface; (b) non-covalent peptide dendrimers (type III), when amino acids are non-covalently linked to the dendrimer framework but encapsulated in the dendrimer cavities [22,63] (**Figure 3.5**).

Similarity to previously described methodologies, the synthesis of PPD dendrimers can also follow a divergent or convergent growth strategy, but using peptide segments as dendrimer core [22]. The synthesis of *L*-lysine dendrimers proposed by *Denkewalter* [60,61,62] was performed using a divergent growth strategy (**Scheme 3.9**). The authors started by reacting *N,N*-di(*tert*-butoxycarbonyl)-*L*-lysine *p*-nitrophenyl ester **22** with 2,6-diamino-*N*-benzyldrylhexanamide (**core**) in DMF (pH between 8 and 9). Then, deprotection of tert-butyl groups **23** was performed with TFA, yielding the desired carbamate derivative **24**. The last step was the reaction of **24** with the amine derivative **22** to give lysine dendrimer **25 G1**. Using this strategy, the authors were able to obtain dendrimers with different sizes (up to **G10** generation).

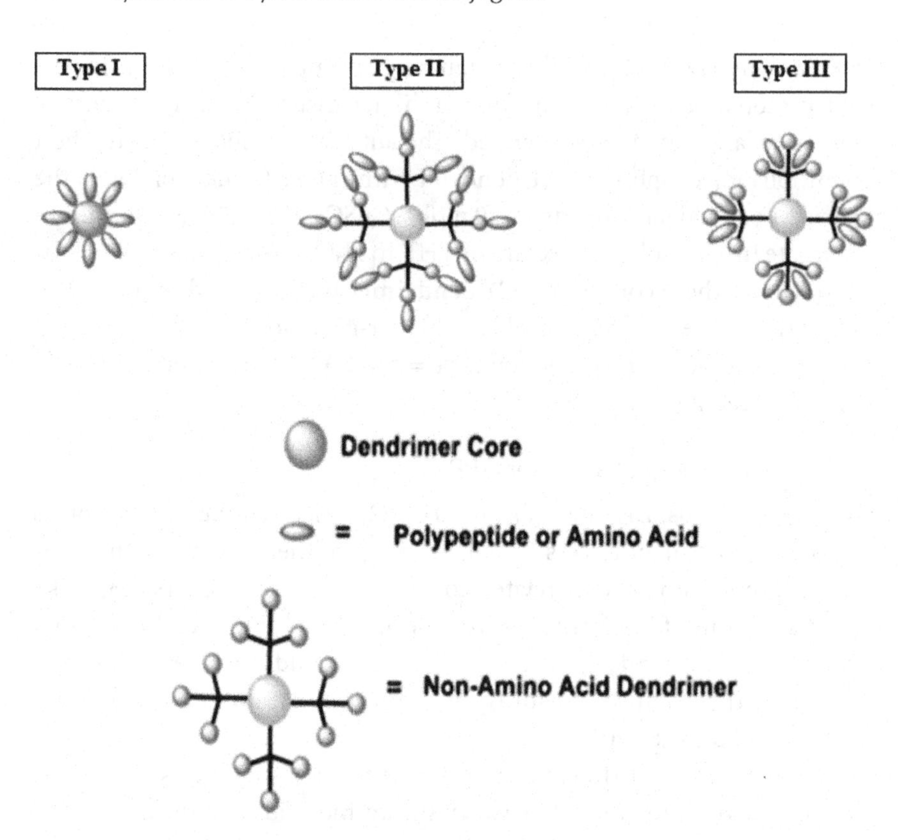

FIGURE 3.5 Schematic representation of PPD types.

Source: Adapted from [63].

Vargas-Berenguel [71,72] reported the synthesis of first-generation amino acid/peptide dendrimers based on a β-CD core *via* convergent synthetic methodology (**Schemes 3.10–3.13**). The author started with the dendron synthesis by converting 3,3'-imino-bispropylamine **26** into its *N*-Cbz-protected derivative **27** through regioselective nucleophilic addition-elimination reaction with benzylchloroformate in the presence of NaOH as base and THF as solvent (**Scheme 3.10a**). Then, protection of the free amine was performed *via* nucleophilic addition-elimination to give the Boc-protected derivative **28 (Scheme 3.10b)**. The next step was the hydrogenolysis of *N*-Cbz from **28** with Pd/C H_2, yielding **29 (Scheme 3.11a)**. Then, **29** reacted with chloroacetic anhydride using Et_3N as base to obtain the desired *N*-chloroacetylated derivative **30 (Scheme 3.11b)**.

Prior to the preparation of the desired **thio-G1-dendron 37 (Scheme 3.13)**, the *N*-mercaptoacetyl **33** was prepared by reacting **31** with

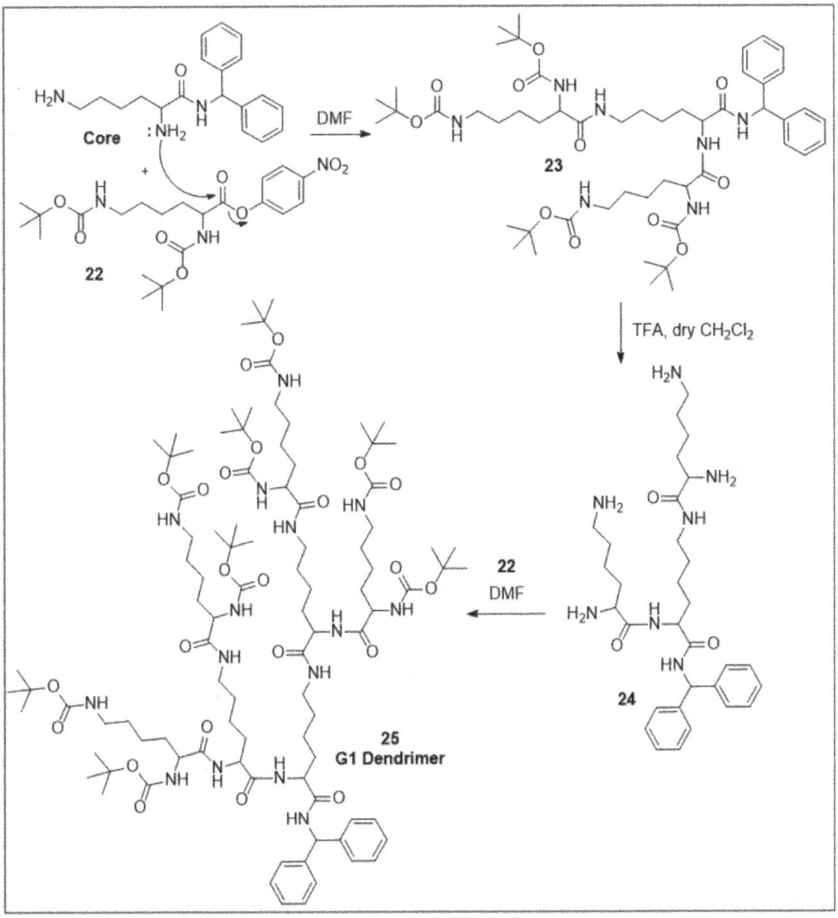

SCHEME 3.9 Synthesis of **G1** polypeptide dendrimer *via* divergent growth strategy.

chloroacetic anhydride, yielding **32**, followed by reaction with thiourea (**Scheme 3.12**). The coupling of the intermediates **30** and **33** was then performed, at room temperature, using Cs_2CO_3 as base and DMF as solvent, to give derivative **34** in 80% yield. The next step in the construction of the dendron was the deprotection (Boc removal) in acidic medium, yielding **35** (**Scheme 3.13a**). This compound reacted with chloroacetic anhydride, giving **36** with 90% yield. Finally, reaction of **36** with thiourea and sodium sulfite allowed for obtaining **thio-G1 dendron 37** with 55% yield (**Scheme 3.13b**).

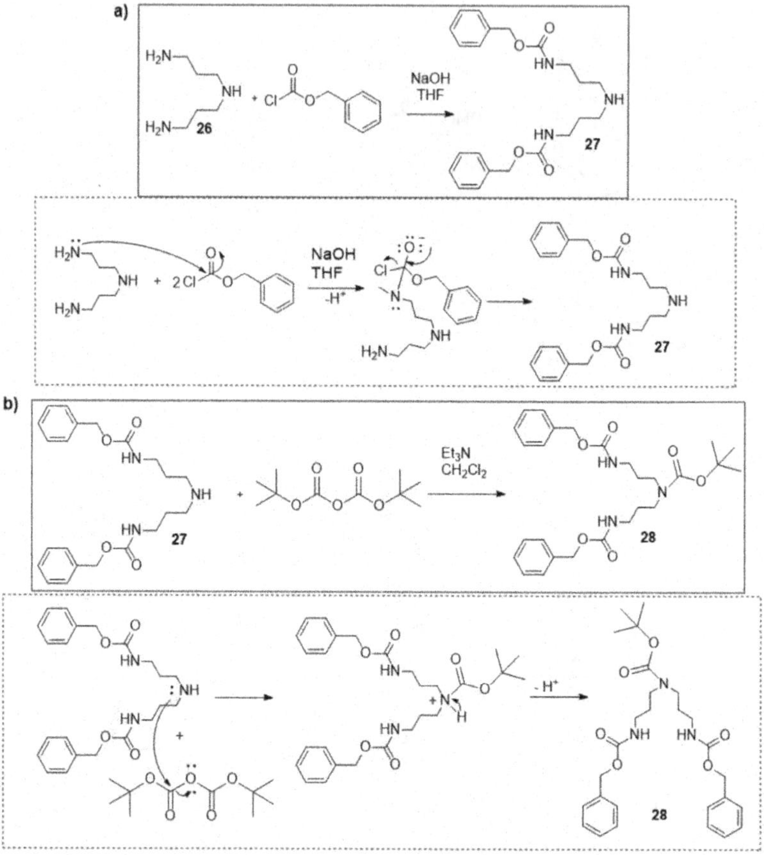

SCHEME 3.10 Synthesis and mechanisms for the preparation of *N*-Boc-protected derivative **28**.

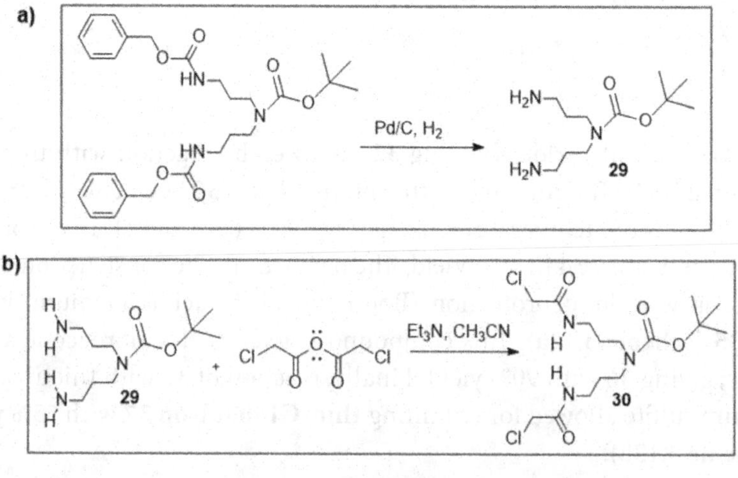

SCHEME 3.11 Synthesis of *N*-chloroacetylated derivative **30**.

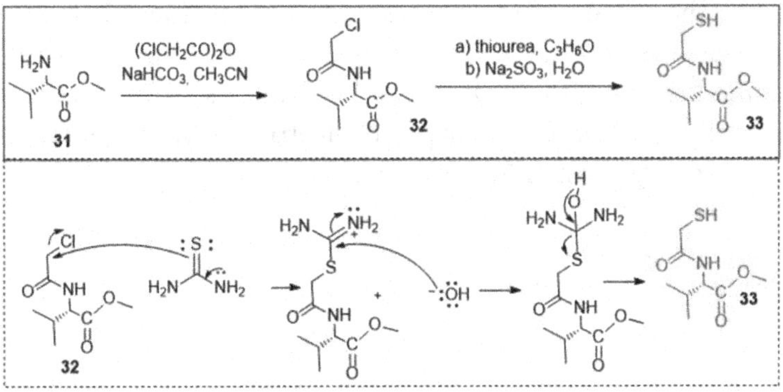

SCHEME 3.12 Synthesis and mechanisms for obtaining *N*-mercaptoacetyl derivative **33**.

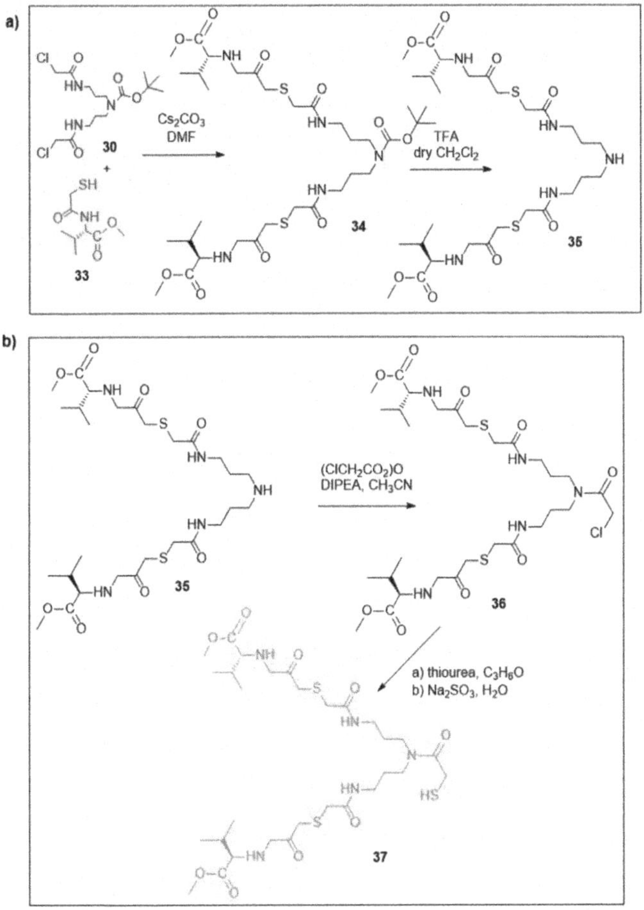

SCHEME 3.13 (a) Synthesis of intermediate **35**; (b) synthesis of **thio-G1 dendron 37**.

Finally, the authors performed the coupling of **thio-G1 dendron 37** and iodinated acetyl-β-cyclodextrin, in seven days, through nucleophilic substitution reaction, using DMF as solvent and Cs_2CO_3 as base (**Scheme 3.14**). Finally, the deprotection of β-cyclodextrin acetyl groups was carried out, giving **G1 dendrimer 38** in 78% yield [71,72].

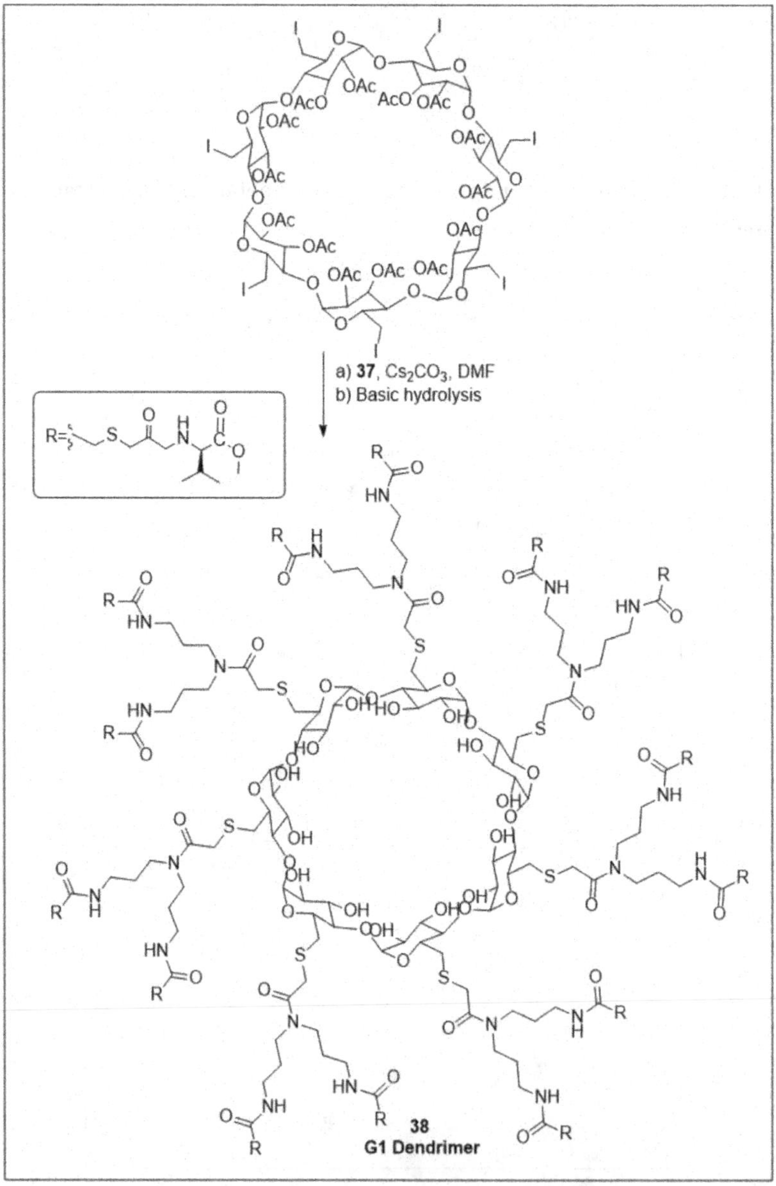

a) **37**, Cs_2CO_3, DMF
b) Basic hydrolysis

38
G1 Dendrimer

SCHEME 3.14 Synthesis of **G1 cyclodextrin-based dendrimer** *via* convergent growth strategy.

3.2 SYNTHETIC METHODOLOGIES FOR PORPHYRIN–DENDRIMER BIOCONJUGATES

3.2.1 Polyester Dendrimers

Moore [73] described the obtaining of poly(phenylester) dendrimers *via* attachment of monodendron esters to the *meta*-positions of the 5,10,15,20-tetrakis(3′,5′-dihydroxyphenyl)porphyrinato manganese (III) chloride (**P43**) using a convergent growing strategy. **P43** was obtained after preparation of the Mn(III) complex of **P2** using manganese (II) chloride salt and DMF as solvent. Then, **PDend 1** dendrimer was obtained by reacting the hydroxylated Mn(III)–metalloporphyrin complex (**P43**) with the corresponding monodendron ester **39**, using DCC as coupling agent and DPTS (4-(dimethylamino)-pyridinium 4-toluenesulfonate) as catalyst (**Scheme 3.15**; see DCC mechanism on **Scheme 3.4**). **PDend1**

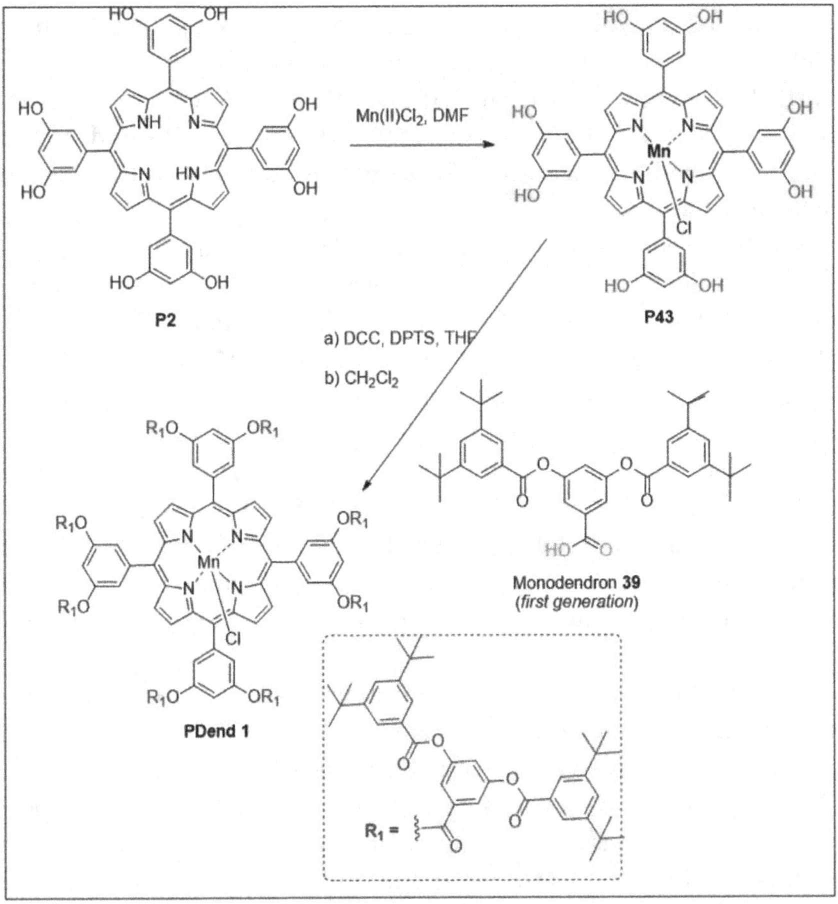

SCHEME 3.15 Synthesis of **G1** dendrimer-Mn(III)-porphyrin **PDend 1** *via* direct DCC coupling reaction.

as obtained in 50% yield, characterized by MALDI-TOF spectrometry, and its purity determined by HPLC (high-performance liquid chromatography). The obtained dendrimer was evaluated as shape-selective oxidation catalysts in the epoxidation of non-conjugated dienes and alkene mixtures of 1-alkene and cyclooctene, showing good results in terms of regioselectivity.

3.2.2 Porphyrin PAMAM Dendrimers

The use of porphyrin PAMAM dendrimers was reported in drug delivery systems, photodynamic therapy, and also gene therapy [74]. *Ramirez-Arroniz* [75], in order to overcome some drawbacks of melphalan (an approved anti-cancer drug), like low water solubility and target specificity, synthesized for the first time water-soluble PAMAM-porphyrin-melphalan bioconjugates. This was achieved via reaction of porphyrin **P2** with methyl bromoacetate and using K_2CO_3 as base and acetone as solvent. Then, the typical sequential reactions, previously described, with ethylenediamine and methyl acrylate were carried out to obtain the dendrimer porphyrin of first (**PDend 2**) and second (**PDend 3**) generations (**Scheme 3.16** mechanisms described in **Scheme 3.1** and **3.5**).

The final step was the covalent linkage of melphalan with PAMAM–porphyrin dendrimers. This was achieved by performing the reaction of **PDend 2** or **PDend 3** with melphalan in methanol under an inert atmosphere, at 45–50°C, for three days (**Scheme 3.17**). The dendrimers were obtained with 77 and 80% yield. After full characterization, the authors studied the *in vitro* anti-cancer activity **PDend 4** and concluded that dendrimers containing melphalan have higher activity in lower doses, against PC-3 cancer cells, than melphalan alone. By comparing both conjugates, it was also possible to conclude that the porphyrin–PAMAM conjugate with 16 melphalan molecules presented the best features, namely, good water solubility and high antiproliferative activity.

Shieh [76] reported the functionalization of PAMAM (**G4**) with porphyrin **P12** and its application as a non-viral carrier in gene therapy. The carboxy group was activated using DCC and *N*-hydroxysuccinimide (NHS), giving the corresponding NHS–porphyrin **P44**. PAMAM–porphyrin **PDend 5** was synthesized *via* conjugation reaction of PAMAM with NHS–porphyrin **P44** using triethylamine as base, in 48 h, at 25°C (**Scheme 3.18** see similar mechanisms on **Scheme 3.4 and 3.5**).

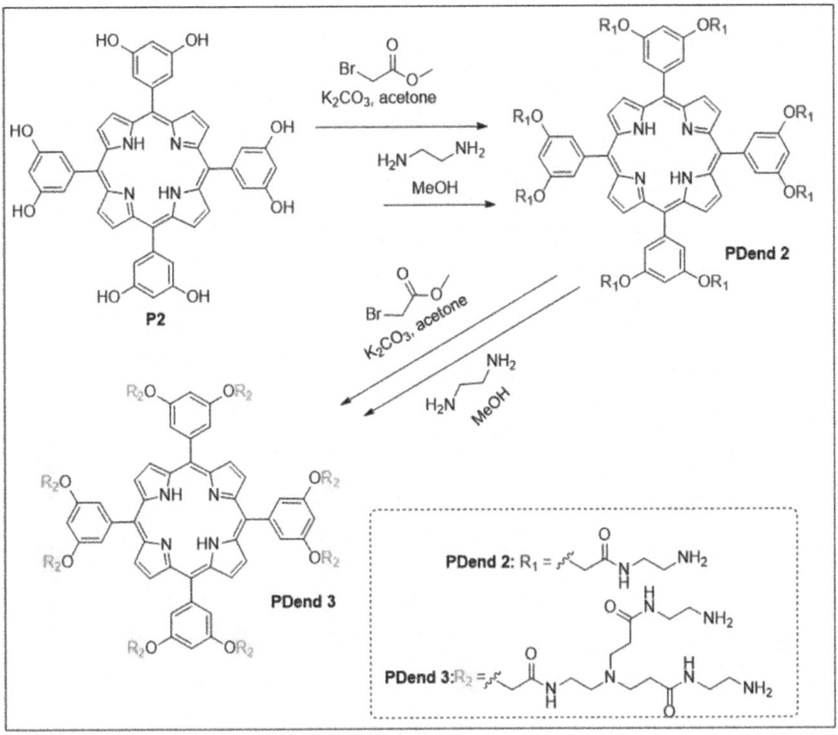

SCHEME 3.16 Synthesis of porphyrin dendrimer **PDend 2** and **PDend 3**.

This bioconjugate **PDend 5** was fully characterized, and its potentiality as gene carrier was evaluated. The authors showed that **PDend 5** has an effective and easy-to-induce gene-transfecting capacity and that it could be a good alternative for *in vivo* photochemical internalization-medicated gene delivery system.

3.2.3 Poly(Peptide) Dendrimers

Odai [77] reported the synthesis of second-, third-, and fourth-generation dendrimer porphyrins. Porphyrin dendrimers were obtained using platinum (II) 5,10,15,20-tetrakis(4-carboxyphenyl)porphyrin (**P5**) as nucleus, poly(*L*-lysine) as dendron, and amino acids (arginine or glutamic acid) as surface (**Scheme 3.19**).

In this study, poly(*L*-lysine)dendron (second, third, and fourth generations) were prepared using commercial *L*-lysine derivatives according to the following: **H-Lys-OMe** and **Boc-Lys(Boc)-OH** reacted in the presence of Et_3N as base, DMF as solvent for 10 min, then with

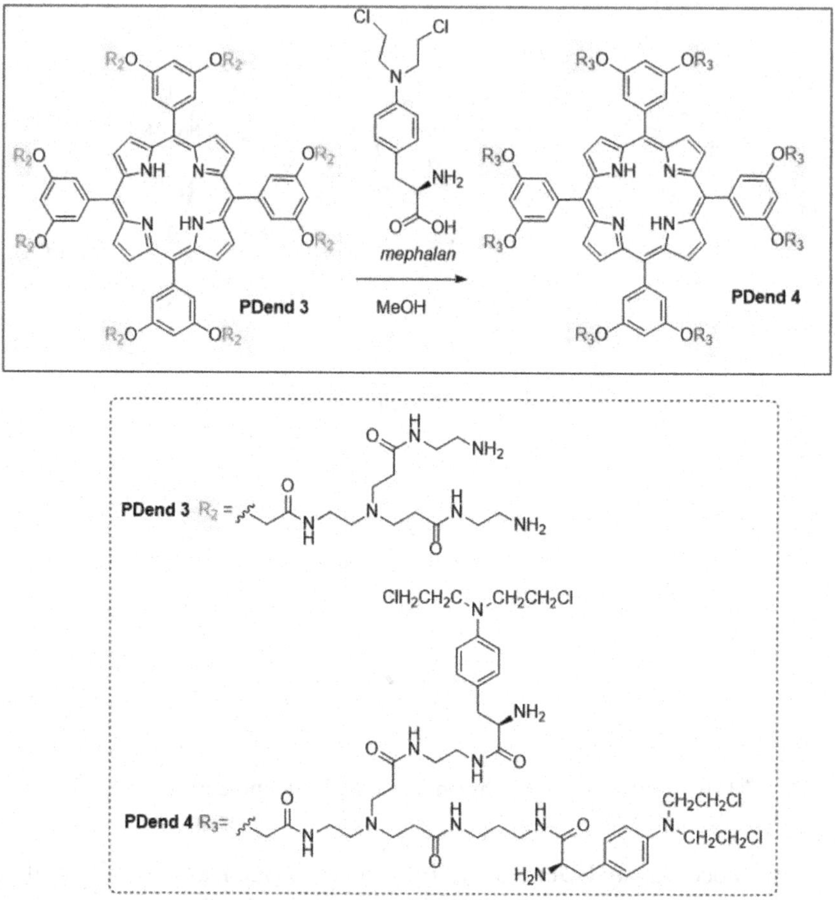

SCHEME 3.17 Coupling reaction of PAMAM–porphyrin **PDend 5** with melphalan.

N,N,N′,N′-tetramethyl-O-(1*H*-benzotriazol-1-yl)uronium hexafluorophosphate (HBTU) and 1-hydroxybenzotriazole hydrate (HOBt.H_2O) for 6 h, at 25°C, yielding second-generation dendron (**MeO-dG2-Boc**) in 92% yield. Then, generation growth of **MeO-dG2-Boc**, until fourth-generation **MeO-dG4-Boc**, was performed as described in **Scheme 3.19**. Dendrimer porphyrin **PDend** was obtained by coupling platinum (II) porphyrin complex **P5** with fourth-generation dendron (**MeO-dG4-Boc**) using OHBt/HBTU.

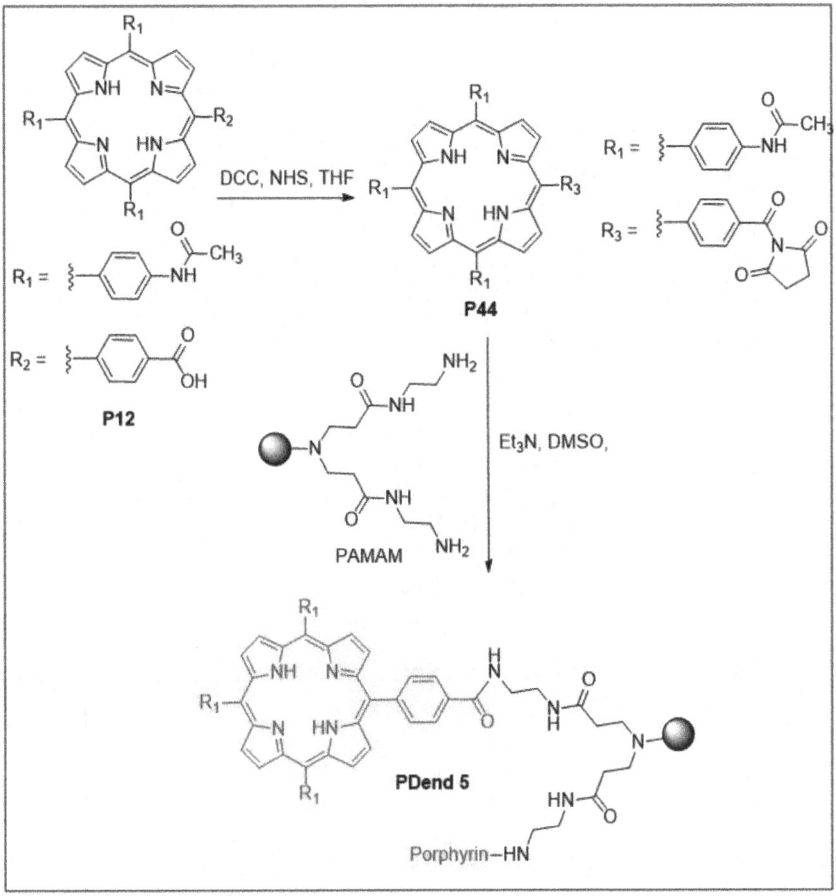

SCHEME 3.18 Synthesis of PAMAM–porphyrin **PDend 5** obtained *via* conjugation of PAMAM and NHS–porphyrin **P44**.

The porphyrin dendrimer (**PDend 6**) structure was confirmed by ^{1}H-NMR and by mass spectrometry. The same authors synthesized similar porphyrin–dendrimer conjugates, but having arginine or glutamic acid as surface groups. All of them showed high potentiality as oxygen sensors for intracellular imaging. It should be highlighted that the spectroscopic features of the dendrimers were not affected by the presence of biological molecules and that high accumulation in cells was observed.

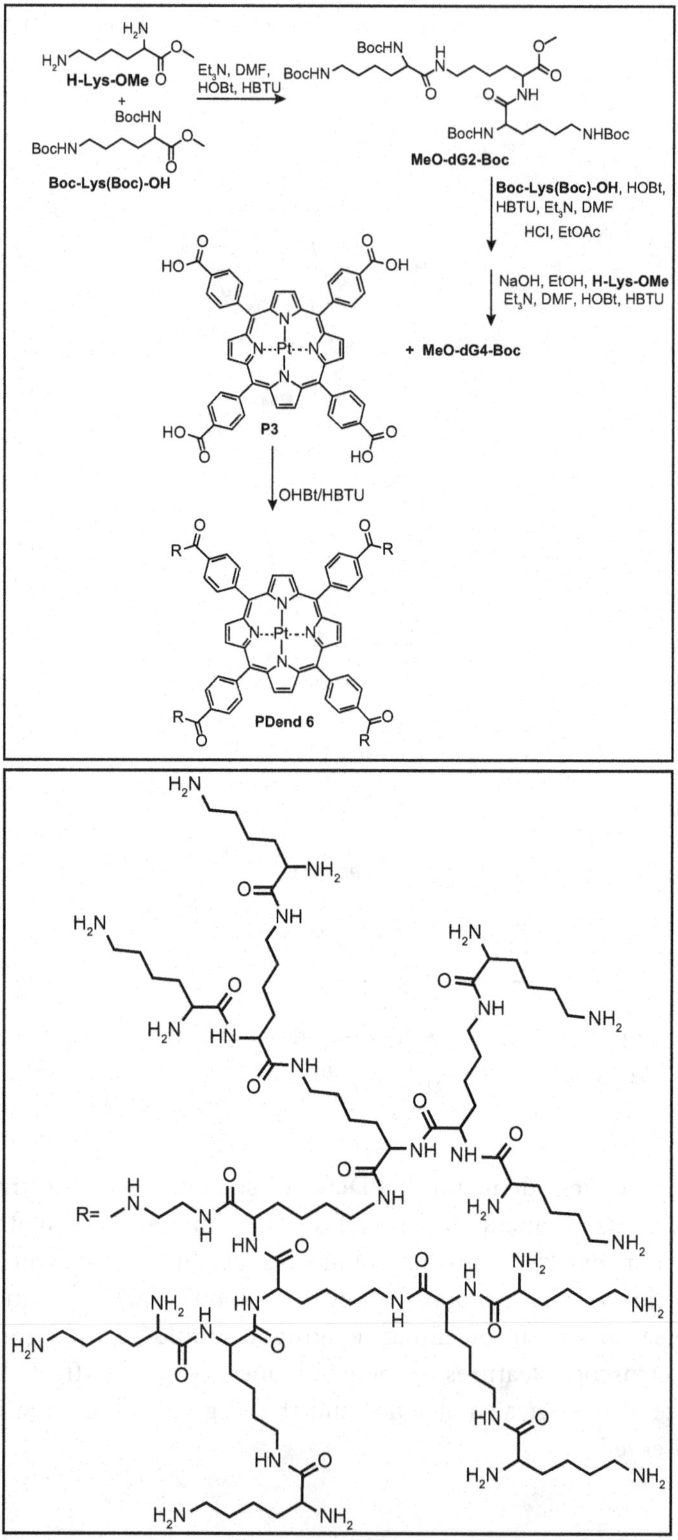

SCHEME 3.19 Synthesis of porphyrin dendrimer **PDend 6**.

REFERENCES

[1] www.fotosearch.com.br/CSP993/k15124114/; accessed in 07 of July 2020.
[2] www.geocaching.com/geocache/GC27332_estuario-do-rio-cavado?guid=9344bb87-c91a-4869-9536-ee66fc06740d; accessed in 07 of July 2020.
[3] www.phillyvoice.com/your-lungs-are-really-amazing-anatomy-professor-explains-why/; accessed in 07 of July 2020.
[4] Buhleier, E., Wehner, W., Vogtle, F., Synthesis, **1978**, 155.
[5] Tomalia, D. A., Baker, H., Dewald, J., Hall, M., Kallos, G., Martin, S., Roeck, J., Ryder, J., Smith, P., *Polym. J.*, **1985**, *17*, 117.
[6] Tomalia, D. A., Baker, H., Dewald, J., Hall, M., Hallos, G., Martin, S., Roeck, J., Ryder, J., Smith, P., *Macromolecules*, **1986**, *19*, 2466.
[7] Astruc, D., Boisselier, E., Ornelas, C., *Chem. Rev.*, **2010**, *110*, 1857.
[8] Caminade, A.-M., Laurent, R., *Coord. Chem. Rev.*, **2019**, *389*, 59.
[9] Yamamoto, K., Imaoka, T., Tanabe, M., Kambe, T., *Chem. Rev.*, **2020**, *120*, 1397.
[10] Sanchez, A., Villalonga, A., Martinez-Garcia, G., Parrado, C., Villalonga, R., *Nanomaterials*, **2019**, *9*, 1745.
[11] Elancheziyan, M., Senthilkumar, S., *App. Surf. Sci.*, **2019**, *495*, 143540.
[12] Scheuble, M., Goll, M., Ludwigs, S., *Macromol. Rapid Commun.*, **2015**, *36*, 115–137.
[13] Gao, W., Wang, J., Lin, Y., Luo, Q., Ma, Y., Dou, J., Tan, H., Ma, C.-Q., Cui, Z., *J. Photochem. Photobiol. A: Chem.*, **2018**, *355*, 350.
[14] Wu, L. P., Ficker, M., Christensen, J. B., Trohopoulos, P. N., Moghimi, S. M., *Bionconjugate. Chem.*, **2015**, *26*, 1198.
[15] Kesharwani, P., Jain, K., Jain, N. K., *Prog. Polym. Sci.*, **2014**, *39*, 268.
[16] Huang, D., Wu, D., *Mater. Sci. Eng. C*, **2018**, *90*, 713.
[17] Donga, Y., Chena, Y., Zhua, D., Shia, K., Maa, C., Zhanga, W., Rocchic, P., Jiangh, L., Liu, X., *J. Control. Release*, **2020**, *322*, 416.
[18] Leiro, V., Garcia, J. P., Tomás, H., Pêgo, A. P., *Bioconjug. Chem.*, **2015**, *26*, 1182.
[19] Battah, S. H., Chee, C.-E., Nakanishi, H., Gerscher, S., MacRobert, A. J., Edwards, C., *Bioconjug. Chem.*, **2001**, *12*, 980.
[20] Mignani, S., Rodrigues, J., Tomás, H., Zablocka, M., Shi, X., Caminade, A.-M., Majoral, J.-P., *Chem. Soc. Rev.*, **2018**, *47*, 514.
[21] Nazemi, A., Gillies, E. R., Dendrimer Bioconjugates: Synthesis and Applications, in *Chemistry of Bioconjugates: Synthesis, Characterization, and Biomedical Applications* (Chapter 5), edited by Narain, R., John Wiley & Sons, Inc., **2014**, 146–182.
[22] Crespo, L., Sanclimens, G., Pons, M., Giralt, E., Royo, M., Albericio, F., *Chem. Rev.*, **2005**, *105*, 1663.
[23] Santos, A., Veiga, F., Figueiras, A., *Materials*, **2020**, *13*, 65.
[24] Boas, U., Christensen, J. B., Heegaard, P. M. H., Dendrimers: Design, Synthesis and Chemical Properties (Chapter 1), in *Dendrimers in Medicine and Biotechnology: New Molecular Tools*, Royal Society of Chemistry, **2006**, ISBN:978-0-85404-852-6.

[25] Bosman, A. W., Janssen, H. M., Meijer, E. W., *Chem. Rev.*, **1999**, *99*, 1665.
[26] Sowinska, M., Urbanczyk-Lipkowska, Z., *New J. Chem.*, **2014**, *38*, 2168.
[27] Hawker, C., Fréchet, J. M. J., *J. Chem. Soc. Chem. Commun.*, **1990**, *15*, 1010.
[28] Hawker, C., Fréchet, J. M. J., *J. Am. Chem. Soc.*, **1990**, *112*, 7638.
[29] Hawker, C., Fréchet, J. M. J., *Macromolecules*, **1990**, *23*, 4726.
[30] Grayson, S. M., Fréchet, J. M. J., *Chem. Rev.*, **2001**, *101*, 3819.
[31] Walter, M. V., Malkoch, M., *Chem. Soc. Rev.*, **2012**, *41*, 4593.
[32] Hawker, C. J., Fréchet, J. M. J., *J. Chem. Soc. Perkin Trans.*, **1992**, *1*, 2459.
[33] Ma, X., Zhou, Z., Jin, E., Sun, Q., Zhang, B., Tang, J., Shen, Y., *Macromolecules*, **2013**, *46*, 37.
[34] Twibanire, J.-d' A., Grindley, T. B., *Polymers*, **2014**, *6*, 179.
[35] Medina, S. H., El-Sayed, M. E. H., *Chem. Rev.*, **2009**, *109*, 3141.
[36] Feliu, N., Walter, M. V., Montanez, M. I., Kunzman, A., Hult, A., Nystrom, A., Malkoch, M., Fadeel, B., *Biomaterials*, **2012**, *33*, 1970.
[37] Gillies, E. R., Dy, E., Fréchet, J. M. J., Szok, F. C., *Mol. Pharmaceut.*, **2005**, *2*, 129.
[38] Jain, K., Kesharwani, P., Gupta, U., Jain, N. K., *Int. J. Pharmaceut.*, **2010**, *394*, 122.
[39] Ihre, H., Hult, A., Soderlind, E., *J. Am. Chem. Soc.*, **1996**, *118*, 6388.
[40] Ihre, H., Jesus, O. L. P., Fréchet, J. M. J., *J. Am. Chem. Soc.*, **2001**, *123*, 5908.
[41] Maiti, P. K., Cagin, T., Wang, G., Goddard III, W. A., *Macromolecules*, **2004**, *37*, 62364.
[42] Wang, Y., Kong, W., Song, Y., Duan, Y., Wang, L., Steinhoff, G., Kong, D., Yu, Y., *Biomacromolecules*, **2009**, *10*, 617.
[43] Shadrack, D. M., Swai, H. S., Munissi, J. J. E., Mubofu, E. B., Nyandoro, S. S., *Molecules*, **2018**, *23*, 1419.
[44] Lyu, Z., Ding, L., Huang, A. Y.-T., Kao, C.-L., Peng, L., *Mater. Today Chem.*, **2019**, *13*, 34.
[45] Majoros, I. J., Myc, A., Thomas, T., Mehta, C. B., Baker, J. R., *Biomacromolecules*, **2006**, *7*, 572.
[46] Tsai, Y.-J., Hu, C.-C., Chu, C.-C., Imae, T., *Biomacromolecules*, **2011**, *12*, 4283.
[47] Dobrovolskaia, M. A., Patri, A. K., Simak, J., Hall, J. B., Semberova, J., Lacerda De Paoli, S. H., McNeil, S. E., *Mol. Pharmaceutics*, **2012**, *9*, 382.
[48] Araújo, R. V., Santos, S. S., Ferreira, E. I., Giarolla, J., *Molecules*, **2018**, *23*, 2849.
[49] Pittelkow, M., Christensen, J. B., *Org. Lett.*, **2005**, *7*, 1295.
[50] Washio, I., Shibasaki, Y., Ueda, M., *Macromolecules*, **2005**, *38*, 2237.
[51] Lee, J. W., Kim, B.-K., Kim, H. J., Han, S. C., Shin, W. S., Jin, S.-H., *Macromolecules*, **2006**, *39*, 2418.
[52] Ito, Y., Higashihara, T., Ueda, M., *Macromolecules*, **2012**, *45*, 4175.
[53] Jishkariani, D., MacDermaid, C. M., Timsin, Y. N., Grama, S., Gillani, S. S., Divar, M., Yadavalli, S. S., Moussodia, R.-O., Leowanawat, P., Camacho, A. M. B., Walter, R., Goulian, M., Klein, M. L., Perec, V., *PNAS*, **2017**, E2275.
[54] Gupta, V., Nayak, S. K., *J. App. Pharm. Sci.*, **2015**, *5*, 117.
[55] Ramstrom, O., The Royal Swedish Academy of Sciences, **2022**.

[56] Agrahari, A. K., Bose, P., Jaiswal, M. K., Rajkhowa, S., Singh, A. S., Hotha, S., Mishra, N., Tiwari, V. K., *Chem. Rev.*, **2021**, *121*, 7638.
[57] Kolb, H. C., Finn, M. G., Sharpless, K. B., *Angew. Chem., Int. Ed. Engl.*, **2001**, *40*, 2004.
[58] Worrell, B. T., Malik, J. A., Fokin, V. V., *Science*, **2013**, *340*, 457.
[59] Scholl, M., Kadlecova, Z., Klok, H. A., *Prog. Polym. Sci.*, **2009**, *34*, 24.
[60] Denkewalter, R. G., Kolc, J. F., Lukasavage, W. J., Macromolecular highly branched homogeneous compound based on lysine units. US Patent 4,289,872. Assigned to Allied Corporation, **1981**.
[61] Denkewalter, R. G, Kolc, J. F., Lukasavage, W. J., Macromolecular highly branched homogeneous compound. US Patent 4,410,688. Assigned to Allied Corporation, **1983**.
[62] Aharoni, S. M., Crosby III, C. R., Walsh, E. K., *Macromolecules*, **1982**, *15*, 1093.
[63] Sapra, R., Verma, R. P., Maurya, G. P., Dhawan, S., Babu, J., Haridas, V., *Chem. Rev.*, **2019**, *119*, 11391.
[64] Okuda, T., Sugiyama, A., Niidome, T., Aoyagi, H., *Biomaterials*, **2004**, *25*, 537.
[65] Sadler, K., Tam, J. P., *Rev. Mol. Biotechnol.*, **2002**, *90*, 195.
[66] Niederhafner, P., Sebestik, J., Jezek, J., *J. Peptide Sci.*, **2005**, *11*, 757.
[67] Cheng, Y., Zhao, L., Li, Y., Xu, T., *Chem. Soc. Rev.*, **2011**, *40*, 2673.
[68] Liu, J., Gray, W. D., Davis, M. E., Luo, Y., *Interface Focus*, **2012**, *2*, 307.
[69] Wan, J., Alewood, P. F., *Angew. Chem. Int. Ed.*, **2016**, *55*, 5124.
[70] Mirakabad, F. S. T., Khoramgah, M. S., Keshavarz, K., Tabarzad, M., Ranjbari, J., *Life Sci.*, **2019**, *233*, 116754.
[71] Caballero-Ortega, F., Gimenez-Martinez, J. J., Garcia-Fuentes, L., Ortiz-Salmeron, E., Gonzalez-Santoyo, F., Vargas-Berenguel, A., *J. Org. Chem.*, **2001**, *66*, 7786.
[72] Muhanna, A. M. A., Ortiz-Salmeron, E., Garcia-Fuentes, L., Gimenez-Martinez, J. J., Vargas-Berenguel, A., *Tetrahedron Lett.*, **2003**, *44*, 6125.
[73] Bhyrappa, P., Young, J. K., Moore, J. S., Suslick, K. S., *JACS*, **1996**, *118*, 5708.
[74] Militello, M. P., Hernandez-Ramirez, R. E., Lijanova, I. V., Previtali, C. M., Bertolotti, S. G., Arbelo, E. M., *J. Photochem. Photobiol. a-Chem.*, **2018**, *353*, 71.
[75] Ramirez-Arroniz, J. C., Klimova, E. M., Pedro-Hernandez, L. D., Organista-Mateos, U., Cortez-Maya, S., Ramirez-Apan, T., Nieto-Cannacho, A., Calderon-Pardo, J., Martinez-Garcia, M., *Drug Dev. Ind. Pharm.*, **2018**, *44*, 1342.
[76] Shieh, M. J., Peng, C. L., Lou, P. J., Chiu, C. H., Tsai, T. Y., Hsu, C. Y., Yeh, C. Y., Lai, P. S., *J. Control. Release*, **2008**, *129*, 200.
[77] Odai, S., Ito, H., Kamachi, T., *J. Clin. Biochem. Nutr.*, **2018**, *65*, 178.

CHAPTER 4

Porphyrin-Based PEG Bioconjugates

4.1 INTRODUCTION

PEGylation, the covalent chemical attachment of polyethylene glycol (PEG) chains to a small biologically active molecule, has emerged as a powerful tool in the field of biotechnology [1,2]. It also contributed to the development of bioconjugates with enhanced therapeutic efficacy, improved pharmacokinetics, and reduced immunogenicity [1,3,4].

Polyethylene glycol (PEG) is a biocompatible synthetic polymer composed of ethylene glycol repeating units. This polymer is generally prepared through the successive ring opening of ethylene glycol epoxide molecules. The reaction begins with the nucleophilic attack of an alkoxide on the epoxide, followed by a sequential opening reaction of the epoxide (**Scheme 4.1**). Typically, the addition of a protic acid increases the reaction rate of the polymerization reaction, giving rise to polymers with controlled molecular weights [5,6].

The introduction of PEG groups into bioconjugates is advantageous because it increases the overall stability by shielding them from proteases and other degradation mechanisms [6,7]. The PEG chain acts like a protective barrier, reducing enzymatic degradation and consequently increasing their *in vivo* circulation time. In addition, PEG bioconjugates typically present long half-life due to reduced renal clearance and concomitant increase in bioavailability. Another relevant property of PEG bioconjugates is its capability to reduce immunogenicity. Indeed, the

DOI: 10.1201/9781003119265-4

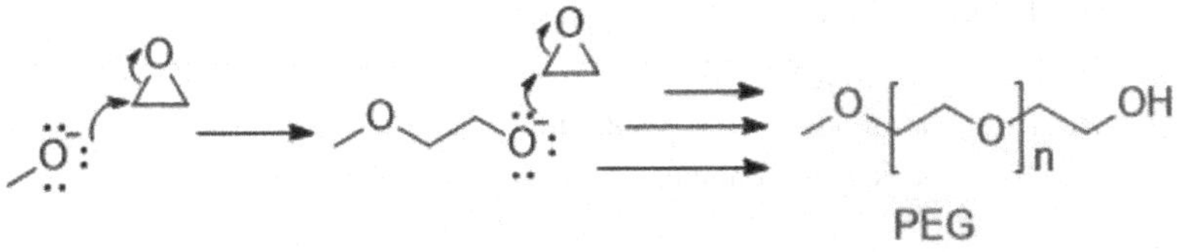

SCHEME 4.1 Mechanism for synthesis of PEG via anionic polymerization.

presence of PEG chains on the surface of a bioconjugate can reduce the recognition of the biomolecule by the immune system, preventing the activation of immune cells and reducing the risk of adverse reactions [5,7]. These properties opened the way for multiple clinical applications of PEG bioconjugates, including in drug delivery, protein engineering, imaging, and diagnostics [8,9,10,11].

Regarding drug delivery, PEGylation has been widely employed by attaching PEG chains to therapeutic molecules, such as small molecules, proteins, or nanoparticles, improving solubility and targeting capabilities, and allowing to modulate controlled release. For instance, PEGylated liposomes have shown great promise in delivering chemotherapeutic agents to tumors while minimizing off-target effects [8].

PEGylation is also used in the field of protein engineering by providing challenging solutions to protein aggregation, stability, and rapid clearance. PEGylated proteins exhibit improved solubility, reduced immunogenicity, and prolonged half-life, making them ideal candidates for therapeutic applications [9].

Another important application of PEG bioconjugates is in the field of medical imaging and diagnostics [12,13]. By coupling imaging agents, such as fluorophores or radionuclides, to PEG chains, it is possible to develop highly sensitive and selective imaging probes. These PEGylated probes can target specific tissues or biomarkers, enabling accurate disease detection and monitoring.

The relevance of PEGgylated bioconjugates is well document by the large number of PEG drugs approved by FDA (Food and Drug Administration) for clinical use. **Table 4.1** summarizes the most relevant ones.

The attachment of PEG to another molecule can be achieved by different chemical strategies, including through direct coupling, activated esters, and functional group–specific reactions [8,14,15,16]. The preparation of PEG bioconjugates typically begins with the conversion of PEG terminal

TABLE 4.1 List of PEG Drugs in Clinical Use

Entry	Drug	Average MW of PEG	Indication	Company	Year Approved
1	*Besremi*	40 kDa	Immunomodulators—used to treat the symptoms of polycythemia vera	PharmaEssentia Corp	2021
2	*Skytrofa*	40 kDa	Growth hormone deficiency	Ascendis	2021
3	*Empaveli*	40 kDa	Paroxysmal nocturnal hemoglobinuria (PNH)	Apellis	2021
4	*Nyvepria*	20 kDa	Neutropenia associated with chemotherapy	Pfizer	2020
5	*Esperoct*	40 kDa	Hemophilia A	Novo Nordisk	2019
6	*Ziextenzo*	20 kDa	Infection during chemotherapy	Sandoz	2019
7	*Udenyca*	20 kDa	Infection during chemotherapy	Coherus BioSciences	2018
8	*Palynziq*	~9 × 20 kDa	Phenylketonuria	BioMarin Pharmaceutical	2018
9	*Revcovi*	80 kDa	Adenosine deaminase deficiency	Leadiant Biosciences	2018
10	*Fulphila*	20 kDa	Infection during chemotherapy	Mylan GmbH	2018
11	*Asparlas*	3139 × 5 kDa	Leukemia	Servier Laboratories	2018
12	*Jivi*	2 × 30 kDa	Hemophilia A	Bayer Healthcare	2017
13	*Rebinyn*	40 kDa	Recombinant coagulation factor IX	Novo Nordisk	2017
14	*Adynovate*	≥1 × 20 kDa	Recombinant antihemophilic factor	Baxalta	2015
15	*Plegridy*	20 kDa	Multiple sclerosis	Biogen	2014
16	*Movantik*	339 Da	Constipation	AstraZeneca	2014
17	*Omontys*	2 × 20 kDa	Anemia	Takeda	2012
18	*Sylatron*	12 kDa	Melanoma	Merck	2011
19	*Krystexxa*	40 × 10 kDa	Gout	Horizon Therapeutics	2010
20	*Cimzia*	40 kDa	Rheumatoid arthritis	UCB	2008
21	*Mircera*	30 kDa	Anemia	Roche	2007
22	*Macugen*	2 × 20 kDa	Macular degeneration	Pfizer	2004
23	*Somavert*	4–6 × 5 kDa	Acromegaly	Pfizer	2003
24	*Neulasta*	20 kDa	Infection during chemotherapy	Amgen	2002
25	*Pegasys*	40 kDa	Hepatitis B and C	Roche	2002
26	*PegIntron*	12 kDa	Hepatitis C, melanoma	Schering	2001
27	*Doxil*	2 kDa	Ovarian cancer, multiple myeloma	Schering	1995
28	*Oncaspar*	69–82 × 5 kDa	Leukemia	Enzon	1994
29	*Adagen*	11–17 × 5 kDa	ADA-SCID	Enzon	1990

* List of drugs containing PEG (number of units) × (molecular weight of each PEG unit).

Note: ADA-SCID: adenosine deaminase severe combined immune deficiency.

hydroxyl group to another functional group with suitable reactivity for subsequent binding to a biologically or pharmacologically active small molecule [16]. **Table 4.2** presents some examples of PEG-functionalized reagents, some of them commercially available, typically used in the synthesis of bioconjugates.

TABLE 4.2 PEG-Reactive Synthons

PEG Reagents	Ref
PEG–C(=O)–CH$_2$–I; PEG–O–S(=O)$_2$–CH$_2$–CF$_3$; PEG–O–(dichlorotriazinyl: N, Cl, Cl); PEG–(CH$_2$)$_n$–Br	[6,7,15,16]
PEG–O–C(O)–O–N-succinimidyl; PEG–O–C(O)–O–N-imidazolyl; PEG–O–C(O)–O–N-benzotriazolyl (N=N); PEG–O–C(O)–O–C$_6$H$_4$–NO$_2$; PEG–O–C(O)–O–C$_6$H$_2$Cl$_3$; PEG–O–C(O)–CH$_2$CH$_2$–C(O)–O–N-succinimidyl	[6,7,16]
PEG–N-maleimide	[6,7]
PEG–S(O)$_2$–CH=CH$_2$	[6,7]
PEG–O–CH$_2$–C≡CH; PEG–N$_3$	[16,17,18]

The next section includes selected examples of porphyrin–PEG bioconjugates using the reagents described in **Table 4.2** and a brief discussion on its medical applications.

4.2 SYNTHETIC METHODOLOGIES FOR PORPHYRIN–PEG BIOCONJUGATION

The synthesis of porphyrin–PEG bioconjugates is one of the most used strategies to enhance biocompatibility and reduce the strong tendency of porphyrin to form aggregates [13,14,15,16,17,18,19,20,21,22,23,24,25].

Pereira [13] reported the synthesis of a PEGylated porphyrin with promising properties for development of a redox MRI probe. The authors started by synthesizing 5,10,15,20-tetrakis(pentafluorophenyl)porphyrin **P9** (**Scheme 4.2**) using the NaY/nitrobenzene method [26]. Then, nucleophilic aromatic substitution in the *para* positions of **P9** was carried out using PEG500 (Mw » 500 g mol^{-1}), DMF, and NaH. The PEGylated derivative **1** was obtained in 65% yield. The following step consisted of

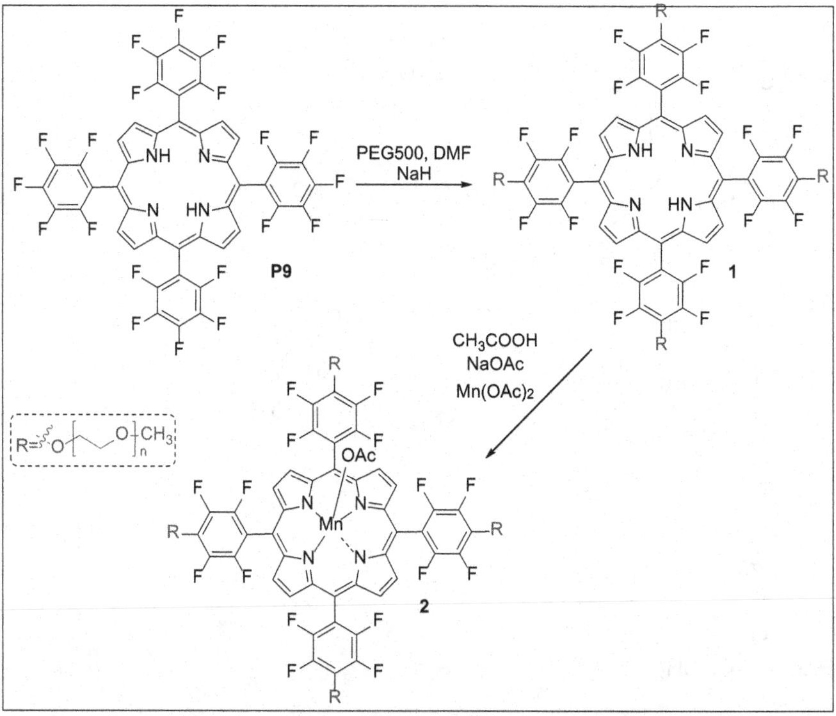

SCHEME 4.2 Synthesis of Mn(III)-5,10,15,20-tetrakis(2,3,5,6-tetrafluoro-4-PEG500-phenyl)porphyrin **2**.

metalation with manganese, using Mn(II) acetate in the presence of acetic acid/sodium acetate, yielding **2** in 85%. This porphyrin was evaluated as potential redox probe by UV-Vis spectroscopy using ascorbic acid as reductant and atmospheric O_2 as oxidant. Additionally, nuclear magnetic relaxation tests indicated that reduction from Mn(III) to Mn(II) led to an increase in relaxivity at low and medium magnetic fields, confirming the potential of porphyrin **2** as redox contrast agent.

Another example of PEGylation of **P9** was described by *Boyle* [19]. In this case, prior to PEGylation, the authors performed bromination of three polyethylenegycol derivatives **3**, with different chain lengths, using PBr_3 **4** as brominating agent, under N_2 at 0°C (**Scheme 4.3a**). The corresponding

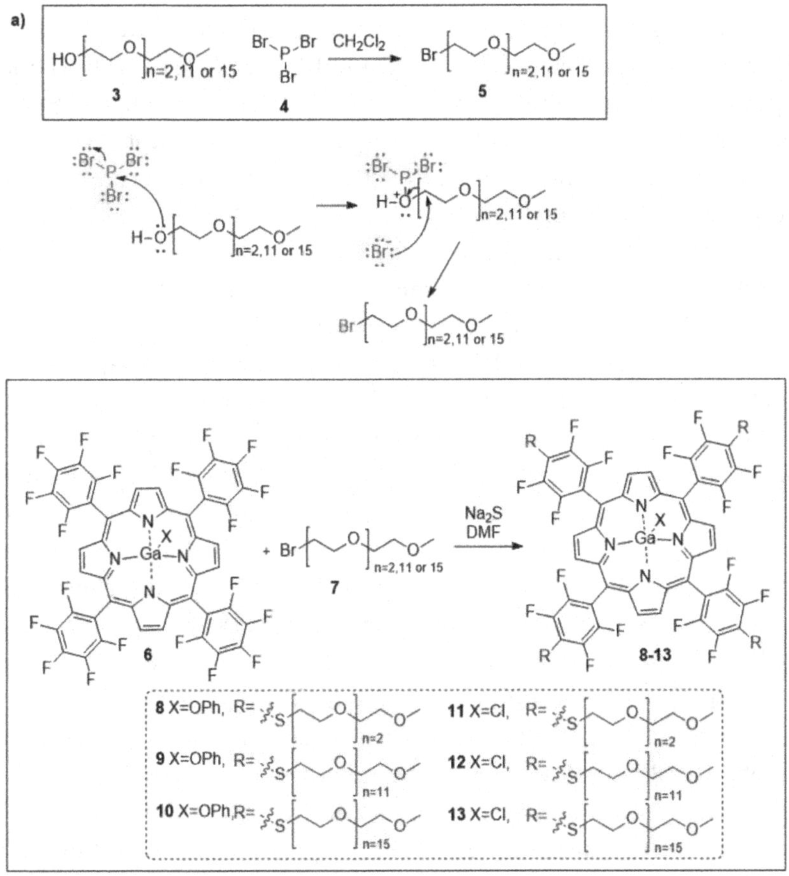

SCHEME 4.3 Synthesis of PEGylated porphyrins (**8–13**): (a) bromination of PEG derivatives; (b) PEGylation of porphyrin derivatives.

brominated PEG derivatives were obtained after work-up, with yields up to 70%. The reaction mechanism is depicted in **Scheme 4.3a**, where first the alcohol functionality is transformed into a good leaving group, with the coordination with the phosphorous atom of PBr_3. Then, a nucleophilic attack of the bromide ion on the electrophilic carbon occurs with the concomitant liberation of the leaving group. PEGylation of free base and Ga (III) complexes of 5,10,15,20-tetrakis(pentafluorophenyl)porphyrin was performed in DMF at room temperature. Here, sodium sulfide was used to promote, prior to PEGylation, the one-pot substitution of PEG bromine by sulfur (**Scheme 4.3b**). After 72 h reaction, all PEG porphyrins were purified *via* extraction procedures or *via* dialysis, and PEGylated porphyrins were obtained with yields >70%. The authors investigated the photodynamic activity of these PEGylated porphyrins against human Caucasian colon adenocarcinoma (Caco2) cells using red light (3.6 J/cm^2 dose). The studies showed that PEGylated porphyrins **12** and **13** that have PEG with 11 and 15 repeating units, respectively, and chlorine as gallium axial ligand were the least phototoxic and the most promising for clinical application.

The synthesis of new short PEG chain–substituted porphyrins using a different synthetic methodology, and its photodynamic activity against cancer cells using light dose of 10 J/cm^2 (lamp wavelength at 425 nm), was reported by *Wierzchowski* (**Scheme 4.4**) [18]. Their strategy involved the previous PEGylation of hydroxy-substituted aldehyde **14** by nucleophilic substitution reaction with brominated PEG **15** (**Scheme 4.4a**). PEGylated aldehyde **16** was therefore obtained after a 24 h reaction at 80°C and purification using silica gel column chromatography in 84% yield. Porphyrin **18** was prepared using the *Adler–Longo* methodology [27] (**Chapter 2**) and purified in silica gel column chromatography. Pure porphyrin **18** was obtained in 28% yield. Then, metalation with zinc acetate (II) was performed, attaining zinc (II) porphyrin **19** in 88% yield (**Scheme 4.4b**). The *in vitro* photodynamic activity of porphyrin **18** and **19** was evaluated against bladder, prostate, and melanoma cancer cells. When compared with other described porphyrins having the PEG group in the *para* rather than *meta* position of the phenyl groups [18], the authors observed that porphyrin **18** and **19** exerted higher phototoxicity against all cell lines. According to the authors, the position of the PEG groups somehow had a crucial effect, increasing phototoxicity if present at *meta* positions [18].

Zhang [20] reported the synthesis of redox stimuli-activated porphyrin photosensitizer based on 5,10,15,20-tetraphenylporphyrin (TPP). The synthetic pathway began with the synthesis of 5,10,15,20-tetraphenylporphyrin

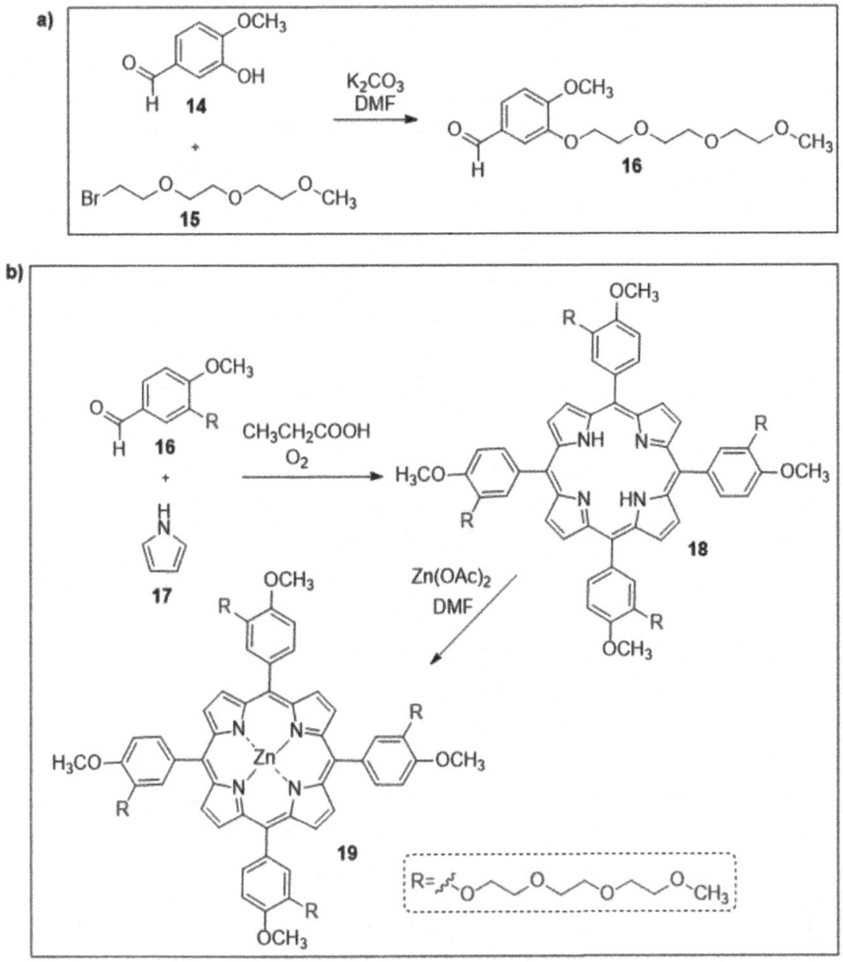

SCHEME 4.4 Synthesis of PEGylated porphyrins **18** and **19**: (a) aldehyde functionalization with PEG; (b) porphyrin **18** and **19** synthesis.

using the *Adler–Longo* methodology [27]. Then, using nitration/reduction strategy [28] (see details in **Chapter 2**), porphyrin **P23** was obtained in 21% yield. To introduce the redox reactive moiety, 2,4-dinitrobenzenesulfonyl, into **P23,** a nucleophilic substitution using 1:1 stoichiometry was performed under N_2 atmosphere, favoring porphyrin mono-substitution, yielding **20** in 87%. The final step involved the coupling of PEG2000COOH to porphyrin **20** using EDC/NHS (1-ethyl-3-(3-dimethylaminopropyl) carbodiimide/succinimidyl ester) coupling strategy, giving porphyrin **21**

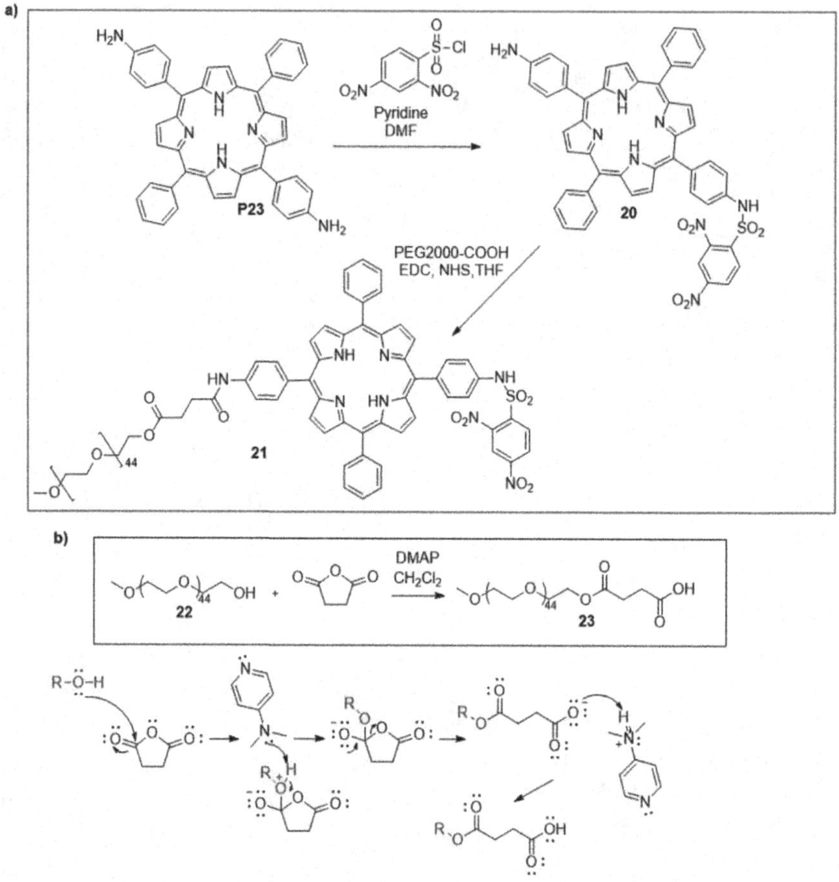

SCHEME 4.5 (a) Synthesis of PEGylated porphyrin derivative **21**; (b) PEG-OH activation and reaction mechanism.

in 85% isolated yield (**Scheme 4.5a**). It is worth mentioning that PEG2000-COOH was used to link the PEG unit to porphyrin **20**, obtained via esterification of PEG2000-OH with succinic anhydride (**Scheme 4.5b**). Mechanistically, the electrophilic carbon of the succinic anhydride first suffers a nucleophilic attack of the PEG hydroxyl group, followed by base deprotonation and ring opening. The final step is protonation, giving the PEG derivative with a pendant carboxylic group.

Furthermore, porphyrin **21** was incorporated into nanomicelles for redox probe evaluation, and it was observed that these composites could dissociate in response to the presence of GSH through porphyrin's sulfonamide group cleavage. This was found to promote an increase both in fluorescence and in the singlet oxygen generation, suggesting that porphyrin

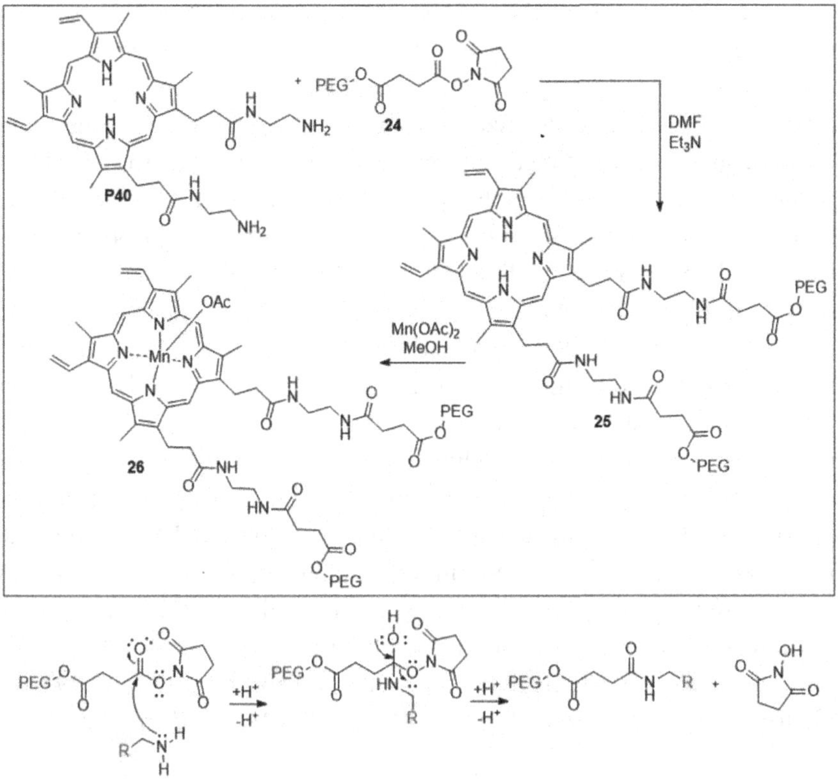

SCHEME 4.6 Synthesis of PEG-protoporphyrin **26**.

21 nanomicelles could act as redox-sensitive photosensitizer, without dark toxicity and phototoxicity.

The use of PEG-succinimidyl as an efficient reagent for porphyrin PEGylation has also been reported. *Sawa* [22] described the synthesis of a PEG-protoporphyrin derivative and evaluated its potential as a catalase mimic and as a drug for the treatment of acetaminophen-induced acute liver failure. The synthetic pathway began with the functionalization of protoporphyrin with amine groups (**P40**; see details in **Chapter 2**), followed by PEGylation with PEG-succinimidyl (**Scheme 4.6**). This reaction was accomplished using DMF and Et_3N at 25°C for 24 h. For purification, bioconjugate **25** was precipitated using cold diethyl ether, and PEG-protoporphyrin **25** was obtained at 98% purity, as determined by HPLC. In the reaction mechanism involved in the PEGylation, the carbonyl group is attacked by the nucleophilic amine, giving a tetrahedral intermediate, followed by elimination of *N*-hydroxysuccinimide. The final step to obtain bioconjugate **26** was metalation with manganese (II) acetate.

In addition to the PEGylation strategies mentioned earlier, another widely used route is the utilization of polyethylene glycol derivatized with a terminal amine group [17,21]. As an example, *Vicente* [21] reported the synthesis of PEG-functionalized porphyrins using a mono-amine polyethylene glycol. The selected synthetic strategy began with the nitration/reduction of 5,10,15,20-tetraphenylporphyrin, as detailed in **Chapter 2**, followed by reaction with glycolic anhydride to obtain the carboxylic acid functionalized–porphyrins **P28–31**. Then, PEGylation with NH_2-PEG **27** of all porphyrin derivatives was performed using the typical coupling peptide strategy HOBt/EDCI (hydroxybenzotriazole/1-ethyl-3-(3-dimethylaminopropyl)carbodiimide) [29] in DMF and Et_3N as base. The reaction was maintained at 25°C for 48 h and after purification by silica gel flash chromatography with suitable eluents (depending on the number of PEG chains introduced). After deprotection with TFA/$CHCl_3$, PEGylated porphyrins **28–31** were obtained in quantitative yields (**Scheme 4.7**). As expected, mono- and di-substituted porphyrins (**28** and **29**) showed high solubility in organic solvents and poor in water, while tri- and tetra-PEGylated forms (**30** and **31**) presented good solubility in both water and organic solvents. All porphyrins were non-toxic in the dark, and just porphyrin **28** and **29** were phototoxic against human HEp2 cells (IC_{50} = 2 mM at 1 J/cm^2 light dose).

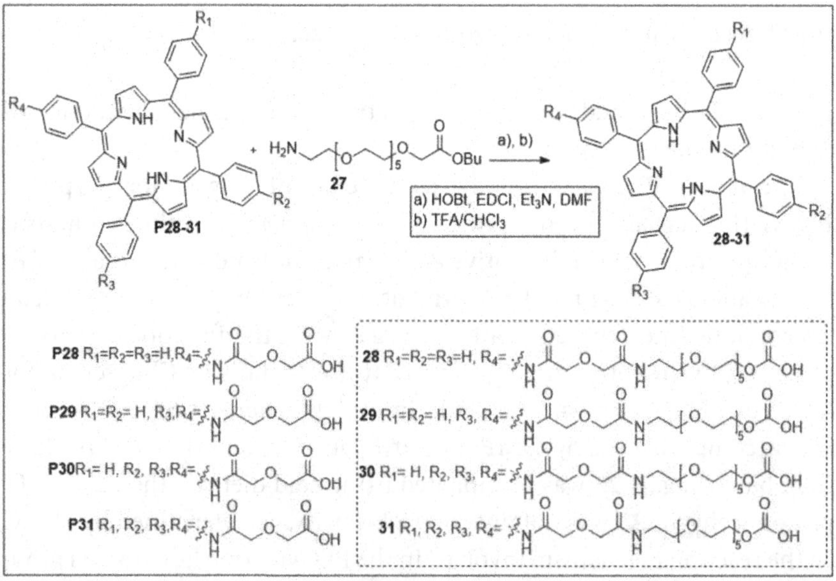

SCHEME 4.7 Synthesis of PEG–porphyrin bioconjugates **28–31**.

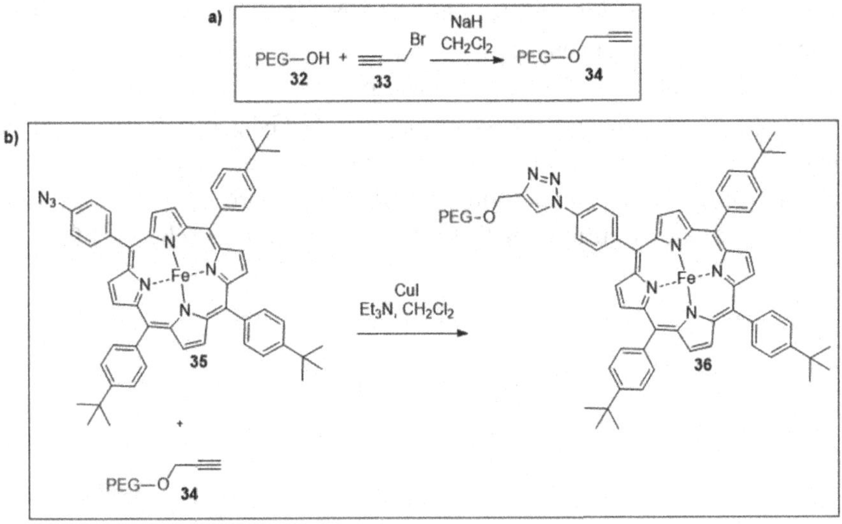

SCHEME 4.8 (a) Synthesis of PEG-alkyne **34**; (b) 1,4-disubstituted 1,2,3-triazole porphyrin **36**.

The authors also observed that these two porphyrins were preferentially found in the mitochondria and the endoplasmic reticulum, while porphyrin **30** and **31** were localized in lysosomes.

There are also a few examples of bioconjugation of PEG with porphyrins via click chemistry using PEG activated with an alkyne or azide functional group [23,24,25]. *Reiser* [23] reported the preparation of a polyethylene glycol iron (II) porphyrin using copper-catalyzed azide-alkyne [3 + 2] cycloaddition click reaction (see mechanism in **Chapter 3**). The authors started by performing propargylation of PEG derivative **32** (**Scheme 4.8a**) with 3-bromoprop-1-yne **33** via bimolecular nucleophilic substitution, using NaH as base and CH_2Cl_2 as solvent, to give the PEG-alkyne derivative **34**. Then, covalent attachment of the PEG alkyne derivative **34** to porphyrin azide **35** (**Scheme 4.8b**) was performed via the typical copper (I) iodide-catalyzed cycloaddition. After total consumption of **34**, addition of an EDTA solution to remove the copper catalyst was performed. Then, after extraction procedures and solvent evaporation, PEG-porphyrin **36** was precipitated out from diethyl ether.

PEG iron (II) porphyrin **36** was used as catalyst for selective olefination of aldehydes with ethyl diazoacetate in the presence of triphenylphosphine. Porphyrin **36** was highly efficient, giving excellent olefin yields (73–99%)

with high selectivity for the *E* isomer. The presence of the PEG5000 covalent attached to the iron porphyrin allowed the authors to recover the catalyst by precipitation/filtration (for ten runs) without relevant activity loss.

Lindsey [24] reported the synthesis of a PEGylated chlorin via click chemistry, having in its structure a water solubilization motif (PEG) and an appropriate functional group handle (phenylpropionic acid group). The synthesis was initiated with the metalation of **C4** (**Scheme 4.9**) with

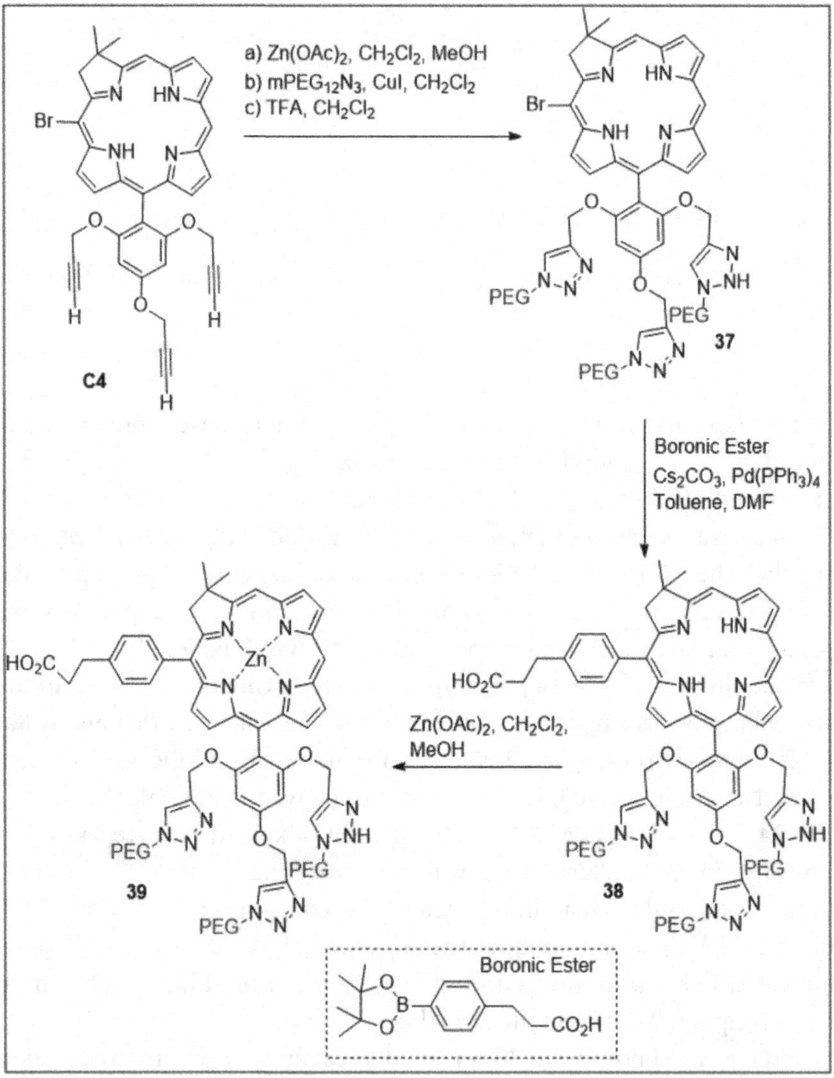

SCHEME 4.9 Synthesis of PEGylated chlorin **39**.

zinc acetate to mask the inner core of the porphyrin. The following step consisted of the azide-alkyne [3+2] cycloaddition click reaction using CuI, sodium ascorbate, and *N,N*-diisopropylethylamine (DIPEA), affording chlorin **37** in 70% yield, after decomplexation in the presence of TFA (**Scheme 4.9**). Then, to obtain the corresponding chlorin, a Suzuki coupling of **37** with 3-(4-(4,4,5,5-tetramethyl-1,3,2-dioxaborolan-2-yl)phenyl)propanoic acid was conducted using toluene/DMF as solvent, Cs_2CO_3 as base, and $Pd(PPh_3)_4$, giving 65% yield of **38**. The last step involved the metalation of **38** with zinc acetate using CH_2Cl_2/MeOH as solvent, for 24 h, at 25°C. The crude product was purified using column chromatography with silica gel as stationary phase, yielding **39** at 65%. To evaluate the potential of **39** as a red fluorophore, the authors performed photophysical characterization and observed that the absorption spectrum exhibited the typical chlorin absorption bands (l_{abs} ~414 and 613 nm) and the emission spectrum a band with a maximum at l_{em} ~616 nm. Also, fluorescence quantum yields were determined in toluene, DMF, and water, giving values of 0.1. All the characterization findings indicated that this chlorin has the features required for a potential red fluorophore probe.

To the best of our knowledge, no PEG–porphyrin bioconjugates are currently in clinical use. Nevertheless, the examples described earlier clearly illustrate the strong potential of such bioconjugates in medicine and catalysis.

REFERENCES

[1] Haag, R., Kratz, F., *Angew. Chem., Int. Ed. Engl.*, **2006**, *45*, 1198.

[2] Duncan, R., Vicent, M. J., *Adv. Drug Deliv. Rev.*, **2013**, *65*, 60.

[3] Liu, S., Maheshwari, R., Kiick, K. L., *Macromolecules*, **2009**, *42*, 3.

[4] Sung, Y. K., Kim, S. W., *Biomater. Res.*, **2020**, *24*, 12.

[5] Shenoi, R. A., Gao, F., Ul-Haq, M. I., Kizhakkedathu, J. N., Bioconjugates Based on Poly(Ethylene Glyol)s and Polyglycerols, in *Chemistry of Bioconjugates: Synthesis, Characterization, and Biomedical Applications*, edited by Narain, R., Wiley-Blackwell, **2014**, 77–103.

[6] Roberts, M. J., Bentley, M. D., Harris, J. M., *Adv. Drug Deliv. Rev.*, **2002**, *54*, 459.

[7] Veronese, F. M., Mero, A., *Biodrugs*, **2008**, *22*, 315.

[8] Kolate, A., Baradia, D., Patil, S., Vhora, I., Kore, G., Misra, A., *J. Control. Release.*, **2014**, *192*, 67.

[9] Zuma, L. K., Gasa, N. L., Makhoba, X. H., Pooe, O. J., *BioMed Res. Int.*, **2022**, 1.

[10] Wang, Z., Ye, Q., Yu, S., Akhavan, B., *Adv. Healthc. Mater.*, **2023**, *12*, 2300105.

[11] Zhou, J., Tong, Y., Zhu, W., Sui, X., Ma, X., Han, C., *Pharm. Dev. Technol.*, **2023**, *28*, 501.

[12] Pinto, S. M. A., Almeida, S. F. F., Tomé, V. A., Prata, A. D., Calvete, M. J. F., Serpa, C., Pereira, M. M., *Dyes Pigm.*, **2021**, *195*, 109677.

[13] Pinto, S. M. A., Calvete, M. J. F., Ghica, M. E., Soler, S., Gallardo, I., Pallier, A., Laranjo, M. B., Cardoso, A. M. S., Castro, M. M. C. A., Brett, C. M. A., Pereira, M. M., Tóth, E., Geraldes, C. F. G. C., *Dalton Trans.*, **2019**, *48*, 3249.
[14] Jevsevar, S., Kunstelj, M., Porekar, G., *Biotechnol. J.*, **2010**, *5*, 113.
[15] Pisal, D. S., Kosloski, M. P., Balu-Iyer, S., *J. Pharm. Sci.*, **2010**, *99*, 2557.
[16] Banerjee, S. S., Aher, N., Patil, R., Khandare, J., *J. Drug Deliv.*, **2012**, *1*.
[17] Chung, C. Y.-S., Fung, S.-K., Tong, K.-C., Wan, P.-K., Lok, C.-N., Huang, Y., Chen, T., Che, C.-M., *Chem. Sci.*, **2017**, *8*, 1942.
[18] Lazewski, D., Kucinska, M., Potapskiy, E., Kuzminska, J., Tezyk, A., Popenda, L., Jurga, S., Teubert, A., Gdaniec, Z., Kujawski, J., Grzyb, K., Pedzinski, T., Murias, M., Wierzchowski, M., *Int. J. Mol. Sci.*, **2022**, *23*, 10029.
[19] Mewis, R. E., Savoie, H., Archibald, S. J., Boyle, R. W., *Photodiagnosis Photodyn. Ther.*, **2009**, *6*, 200.
[20] Xue, Y., Tian, J., Liu, Z., Chen, J., Wu, M., Shen, Y., Zhang, W., *Biomacromolecules*, **2019**, *20*, 2796.
[21] Sibrian-Vazquez, M., Jensen, T. J., Vicente, M. G. H., *J. Photochem. Photobiol. B: Biology*, **2007**, *86*, 9.
[22] Zhang, T., Fang, J., Tsutsuki, H., Ono, K., Islam, W., Sawa, T., *Biol. Pharm. Bull.*, **2019**, *42*, 1199.
[23] Chinnusamy, T., Rodionov, V., Kuhn, F. E., Reiser, O., *Adv. Synth. Catal.*, **2012**, *354*, 1827.
[24] Matsumoto, N., Taniguchi, M., Lindsey, J. S., *J. Porphyr. Phthalocyanines*, **2020**, *23*, 3–17.
[25] Dumoulin, F., Ahsen, V., *J. Porphyr. Phthalocyanines*, **2011**, *15*, 482.
[26] Silva, M., Fernandes, A., Bebiano, S. S., Calvete, M. J. F., Ribeiro, M. F., Burrows, H. D., Pereira, M. M., *Chem. Comm.*, **2014**, *50*, 6571.
[27] Adler, A. D., Longo, F. R., Finarelli, J. D., Goldmacher, J., Assour, J., Korsakoff, L., *J. Org. Chem.*, **1967**, *32*, 476.
[28] Luguya, R., Jaquinod, L., Fronczek, F. R., Vicente, M. G. H., Smith, K. M., *Tetrahedron*, **2004**, *60*, 2757.
[29] Sibrian-Vazquez, M., Jensen, T. J., Hammer, R. P., Vicente, M. G. H., *J. Med. Chem.*, **2006**, *49*, 1364.

CHAPTER 5

Porphyrin-Based Carbohydrate Bioconjugates

5.1 INTRODUCTION

Carbohydrates are one of the most abundant organic-based compounds found in the biosphere, containing atoms of carbon, hydrogen, oxygen, and in some cases, nitrogen and phosphorus [1,2]. This class of biomolecules has an empirical formula of $(CH_2O)_n$, $(n \geq 3)$, being the main energy source for animals and plants [3]. Carbohydrates are divided into three main classes: monosaccharides, oligosaccharides, and polysaccharides. The first class includes trioses ($C_3H_6O_3$), tetroses ($C_4H_8O_4$), pentoses ($C_5H_{10}O_5$), hexoses ($C_6H_{12}O_6$), and heptoses ($C_7H_{14}O_7$), of which hexoses, namely, glucose and fructose, have been the most widely used molecules for the preparation of bioconjugates. Oligosaccharides consist of multiple covalently linked monosaccharides, typically having two to ten units, including lactose and maltose as examples, while polysaccharides (glycans) are biopolymers containing monosaccharides linked by a glycoside bond, such as chitosan, heparin, glycans, starch, cellulose, glycogen, pectin, hyaluronic acid, and lignin [4,5] (**Figure 5.1**).

Due to their essential role in *in vivo* biochemical processes, there is a significant interest in the development of carbohydrate-based bioconjugates [6,7]. The bioconjugation of target molecules such as drugs, diagnostic

DOI: 10.1201/9781003119265-5

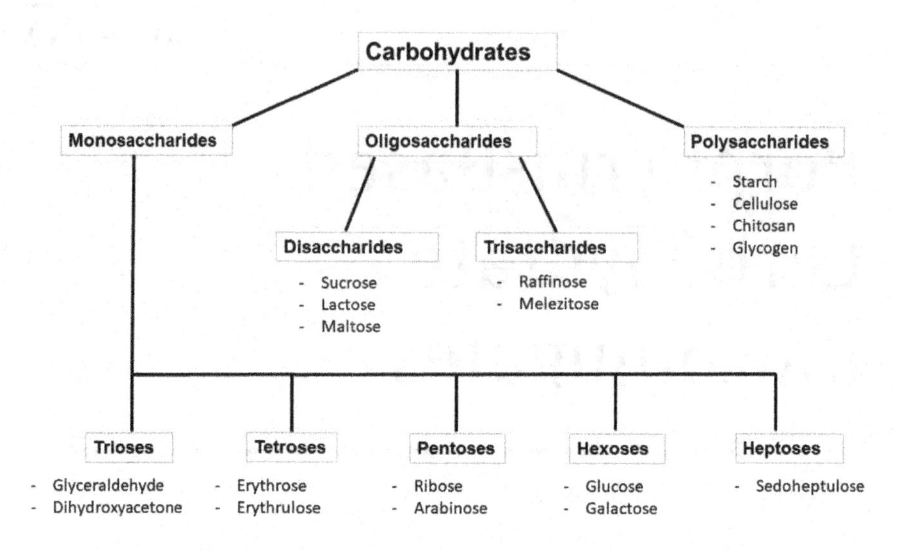

FIGURE 5.1 General classification of the carbohydrates.

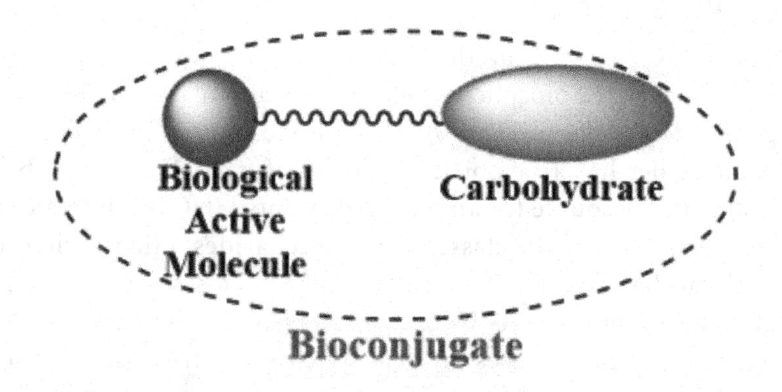

FIGURE 5.2 General structure of a carbohydrate-based bioconjugate.

agents, and antigens with carbohydrates can modulate their therapeutic efficiency by improving solubility, bioavailability, stability, *in vivo* half-life, and delivery to targeted organs or tissues [8] (**Figure 5.2**). There are many studies in the literature that clearly demonstrate the high potential of combining carbohydrates with small biologically active chemical entities, as illustrated in **Figure 5.3** [9,10,11].

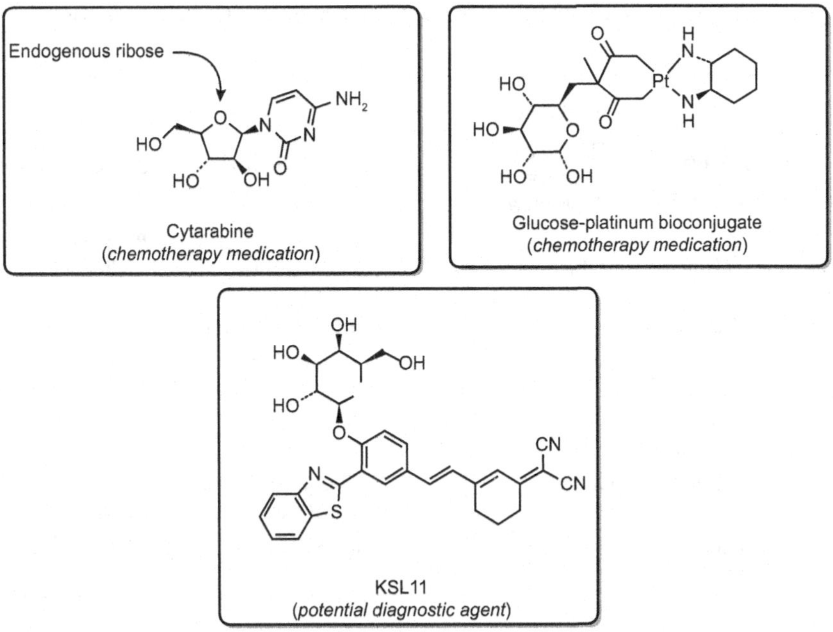

FIGURE 5.3 Examples of carbohydrate bioconjugates containing a drug in clinical use [9,10,11].

5.2 GENERAL SYNTHETIC METHODS FOR BIOCONJUGATION OF PORPHYRINS WITH CARBOHYDRATES

The development of modern organic chemical processes presents new perspectives for synthesizing carbohydrate-based bioconjugates [12] specifically through click chemistry, aromatic nucleophilic substitution, amidation, and reductive amination (**Table 5.1**).

The selection of a suitable synthetic methodology for preparing bioconjugates incorporating carbohydrates into their structures depends, as highlighted in earlier chapters, on the molecular structure/functional groups present in each of the entities involved. For example, click chemistry is a widely used strategy for this purpose. As mentioned in **Chapter 3**, it requires specific conditions, such as Cu(I) as catalyst, to improve the selectivity of the 1,3-dipolar cycloaddition between terminal alkynes and azides [13] (**Table 5.1,** entry 1).

TABLE 5.1 Main Synthetic Methodologies for the Synthesis of Carbohydrate-Based Bioconjugates

R_1 + R_2 → R_3

Carbohydrate   Biological Active Molecule   Carbohydrate-based conjugates

Entry	Method	R_1	R_2	Reagents	Reaction Conditions	Ref
1	Click chemistry	N_3	—C≡CH	DIPEA; CuI 0.1 M NaOH (aq) (2 steps)	-	[13]
2	Aromatic nucleophilic substitution	OH	Halogen (weak base)	Base (e.g., NaH, *t*-BuOK, and NaOH)	-	[14]
3	Amidation	COOEt	NH_2	$Na_2B_4O_7$, $KHCO_3$ (aq) (1 step)	pH: 9.05–9.30	[15]
4	Reductive amination	COH	NH_2	$NaCNBH_3$ (1 step)	pH: 7.0; 37°C	[16]

Additionally, the nucleophilic aromatic substitution strategy involves a carbohydrate (bearing -OH groups) as nucleophile, a biologically active molecule with a good leaving group (**Table 5.1,** entry 2), and a base to activate the hydroxyl group [14]. Another relevant approach involves the formation of an amide linkage by activating the -COOH functional group with coupling agents, such as dicyclohexylcarbodiimide (DCC), 1-(3-dimethylaminopropyl)-3-ethylcarbodiimide hydrochloride (EDCI). Lastly, it is worth mentioning that the reductive amidation strategy is also widely used to obtain carbohydrate bioconjugates, and it involves a reaction between a carbohydrate bearing a formyl group with an amine-functionalized biologically active molecule [15]. This procedure requires a reducing agent, such as sodium cyanoborohydride ($NaCNBH_3$), to facilitate the conversion of the imine (intermediate) into the corresponding amine compound [16] (**Table 5.1,** entry 4).

Carbohydrate–porphyrin bioconjugates can also be directly obtained through the condensation of carbohydrate-functionalized aldehydes with pyrrole. **Chapter 2** provides a detailed description of the methodologies for obtaining these porphyrins. So the following section will present selected examples of porphyrin–carbohydrate bioconjugates, including reaction conditions, reagents, and applications, which will also be presented.

5.3 LINKAGE OF CARBOHYDRATES WITH PORPHYRIN-BASED COMPOUNDS

Dixon [17] synthesized a series of carbohydrate–porphyrin bioconjugates bearing various sugars (e.g., Gal, Glc, GalNAc, GlcNAc, Lac) and spacers of different lengths using click chemistry as synthetic strategy. These carbohydrate-based bioconjugates were synthesized by catalytic alkoxycarbonylation reaction between Zn(II)-10-bromo-5,15-diphenylporphyrin derivative **1** and TMS-protected alkynol spacers (2–6 carbons) **2–6** in the presence of $Pd(OAc)_2$ as Pd(II) salt, xantphos as ligand, and Cs_2CO_3 as base, under N_2 atmosphere (**Scheme 5.1**). Then, removal of TMS protecting group was carried out using tetrabutylammonium fluoride (TBAF), at room temperature, for 15 min. The deprotected Zn(II)-porphyrin derivatives **7–11** were purified by automated chromatography and fully characterized. Carbohydrate-based bioconjugates **12–16** were further obtained by reacting **7–11** with carbohydrate derivatives such as Gal, Glc, GalNAc,

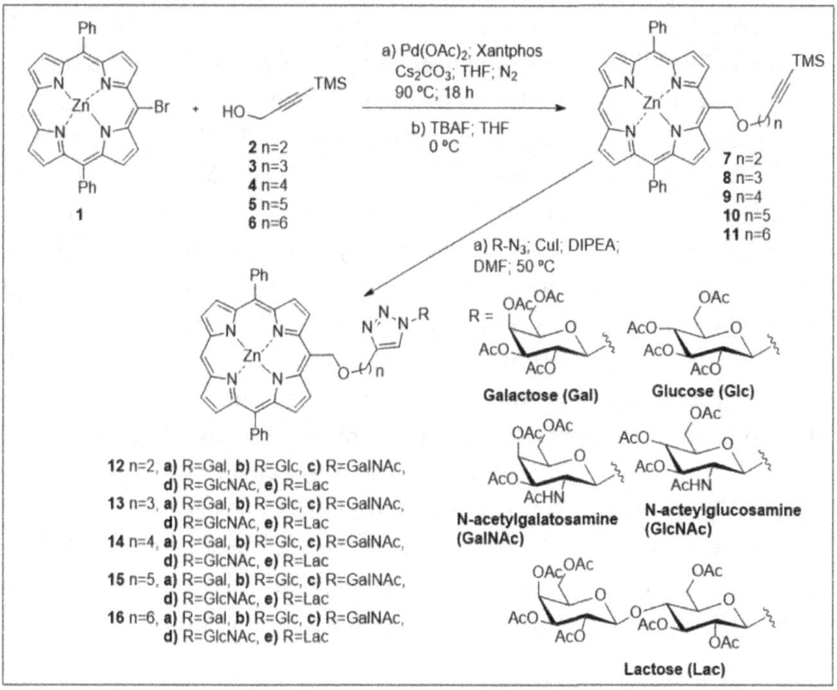

SCHEME 5.1 General synthetic method for preparation of carbohydrate-based bioconjugates using Zn(II)-porphyrin derivative **1** and azide sugar derivatives by click chemistry approach.

GlcNAc, and Lac using CuI as catalyst, DIPEA as base, and DMF as solvent. The reaction was stirred at 50°C for 48 h, and after purification by flash chromatography, the carbohydrate-based bioconjugates were obtained in yields up to 90% (**Scheme 5.1**). aPDT efficiency of carbohydrate-based porphyrins **12–16** was evaluated against *M. smeg*, and it was observed that the lower-length linkers increase the photodynamic efficiency.

Rosa reported an efficient synthetic methodology for galactoporphyrin derivatives [14] involving an aromatic nucleophilic substitution reaction between nitroporphyrin derivatives **17–24** and 3,4-di-*O*-isopropylidene-α-D-galactopyranose **25**. The authors evaluated several base/solvents and concluded that, at 65°C, *t*-BuOK (as base) and Et_3N (as solvent) are the best reagents to obtain the bioconjugates **26–33** in yields up to 78%. While this methodology proves useful for glycoporphyrin synthesis *via* S_NAr methodology, a high excess of base is necessary (**Scheme 5.2**). Regarding the mechanism, the presence of electron-withdrawing groups such as -NO_2 on the phenyl groups of the porphyrin makes it susceptible to nucleophilic attack by the hydroxyl anion of the carbohydrate. Galactoporphyrin derivatives **26–33** exhibited lower cytotoxicity and increased uptake by A549 and MCF-7 cell lines when compared to the unglycosylated derivative.

Hirohara [18] reported the synthesis of a glycoconjugated pyrrolidine-fused chlorin **34** (**Scheme 5.3**). *S*-glycosylation of Pd(II)-chlorin **34** with 2,3,4,6-tetra-*O*-acetyl-1-thio-*β*-D-glucopyranoside **35** was carried out using DMF as solvent and triethylamine as base, at room temperature, for 12 h. The crude product was purified by silica gel column chromatography, followed by preparative GPC (gel permeation chromatography), yielding **36** in 66% yield. The last step comprised the removal of the acetyl groups using sodium methoxide as base and a mixture of CH_2Cl_2/MeOH as solvent, followed by purification using reverse-phase HPLC and dialysis. Glycoconjugated pyrrolidine-fused chlorin **37** was obtained in 59% yield. Aiming to assess the *in vivo* photocytotoxicity of the synthesized chlorin **37**, the authors performed phototoxicity studies using HeLa cell line and metastatic B16-BL6 melanoma cells. They observed that **11** demonstrated excellent phototoxicity effect, despite displaying a moderate f_D (0.28). This suggests the involvement of both type I and type II mechanisms [19].

As previously mentioned, the amidation reaction is frequently applied for the synthesis of carbohydrate-based bioconjugates. *Stasio* [19] reported the coupling reaction between a non-symmetric porphyrin **P13** bearing a carboxylic acid functional group with *O*-acetylated glucosamine **38** (**Scheme 5.4**)

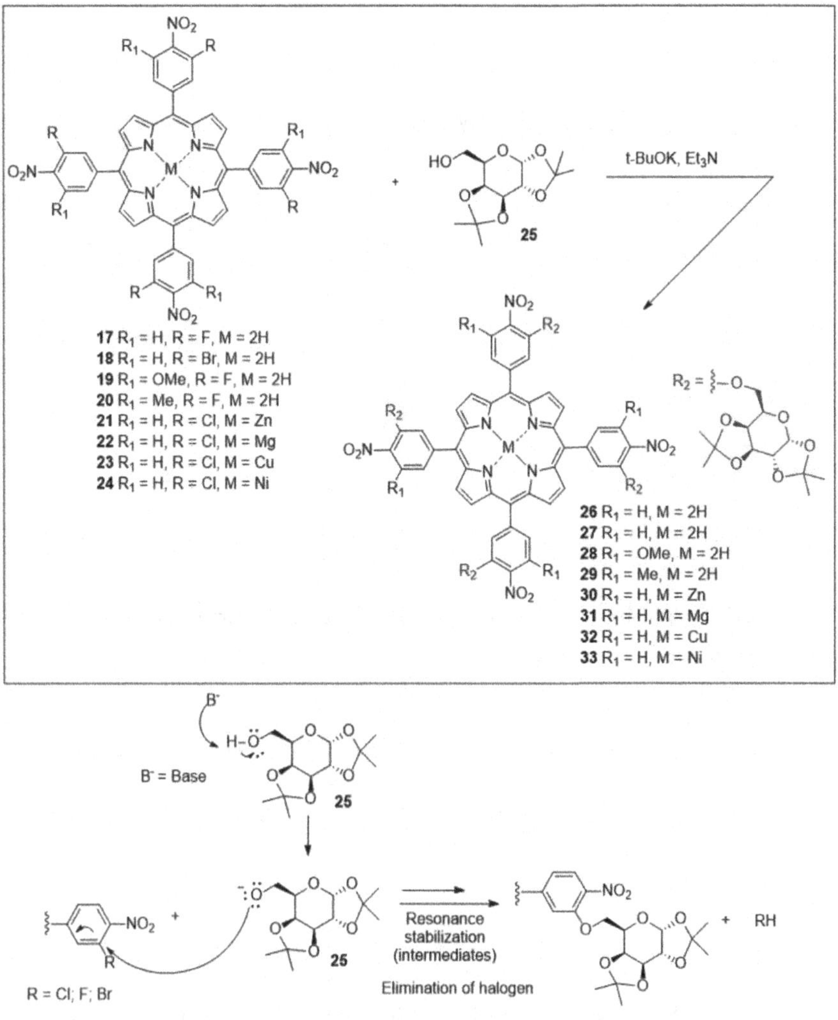

SCHEME 5.2 Synthesis of galactoporphyrin derivatives *via* aromatic nucleophilic substitution reaction.

in the presence of 1-(3-dimethylaminopropyl)-3-ethylcarbodiimide hydrochloride (EDCI) and using mixture of CH_2Cl_2/DMF as solvent, at 0°C (mechanism described in **Chapter 3**). After 24 h reaction, the obtained bioconjugate **39** was purified using silica gel chromatography, affording **39** in 65% yield. The authors observed that bioconjugate **39** showed high cellular uptake in human colorectal adenocarcinoma cells (HT29) and cellular accumulation in the endoplasmic reticulum than porphyrin **P13**.

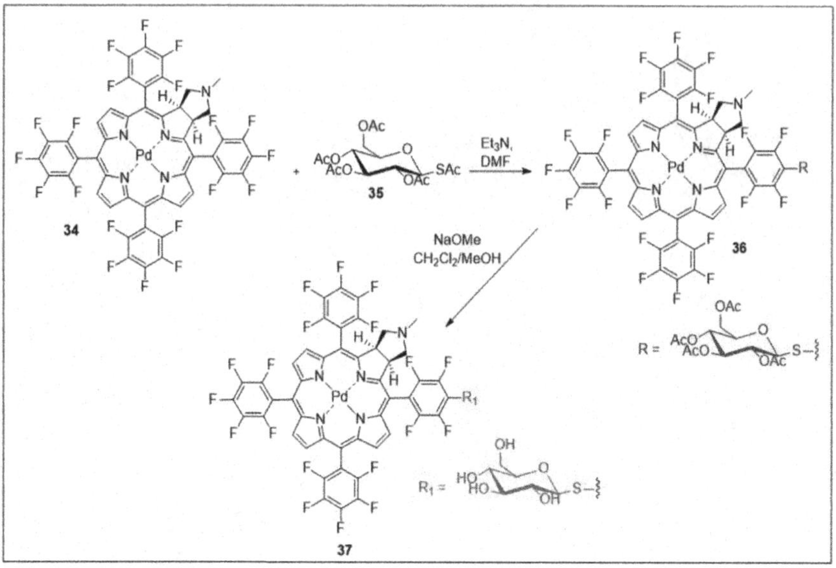

SCHEME 5.3 Synthesis of pyrrolidine-fused Pd(II)-chlorin-carbohydrate bioconjugate **37**.

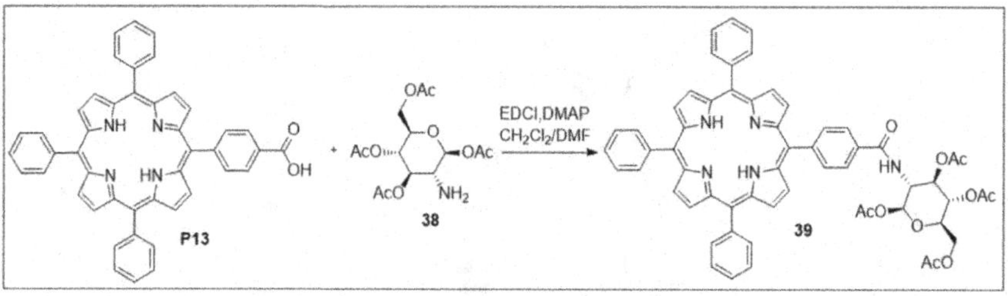

SCHEME 5.4 Synthesis of 5-[4-(1,3,4,6-tetra-*O*-acetyl-2-amido-2-desoxy-*β*-D-glucopyranose phenyl)]-10,15,20-triphenyl porphyrin.

Asayama described another example of a carbohydrate-based bioconjugate (**41**) obtained *via* reductive amination (**Scheme 5.5**) [20]. The water-soluble porphyrin derivative **P7**, bearing an amine group, reacted with lactose monohydrate **40** for three days, at 40°C, using sodium borate buffer (at pH 8.5) as solvent. Then, $NaBH_3CN$ reducing agent was added, and the reaction was kept another four days, at 40°C. The obtained compound was purified using a cellulose ester membrane. The synthesized bioconjugate **41** showed significant superoxide dismutase (SOD) activity and low cellular cytotoxicity. Also, the presence of lactose at **41** enhanced its

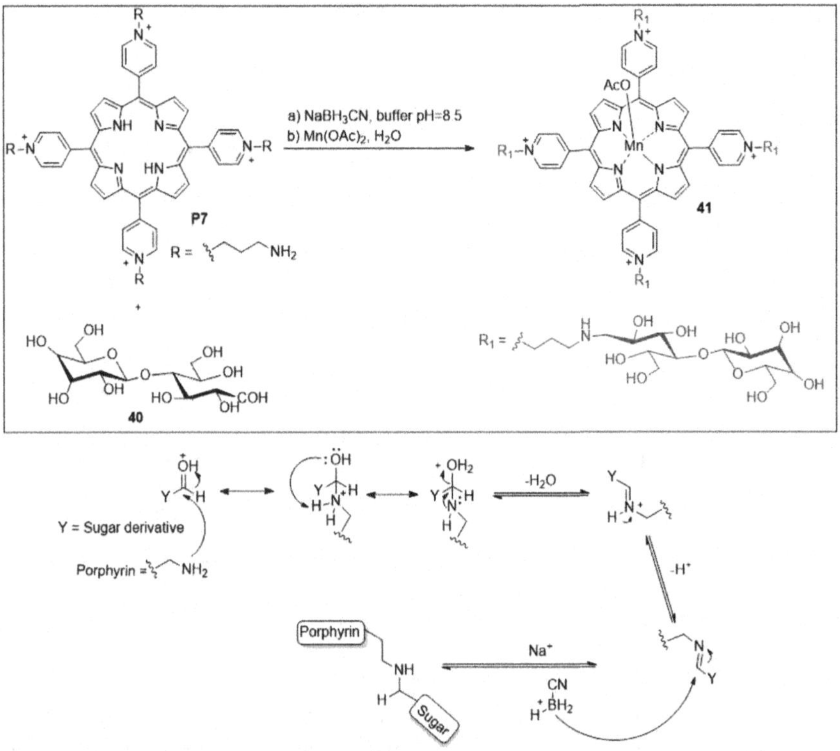

SCHEME 5.5 Application of reductive amination approach for synthesis of Mn-porphyrin-lactose (4-*O*-a-D-galactopyranosyl- D-glucose).

recognition by human hepatoma HepG2 cells when compared with other metalloporphyrins lacking this carbohydrate units.

Vedachalam [21] described the synthesis of a porphyrin carbohydrate-based bioconjugate obtained directly during the porphyrin synthesis process. Firstly, *C*-glycosyl dipyrromethane **44** is obtained by reaction of pyrrole with the carbohydrate **42** (5:1 molar ratio) using $BF_3 \cdot Et_2O$ as catalyst (**Scheme 5.6a**). The condensation of **44** with different aldehydes to give **53–60** was performed using the well-known *Lindsey* methodology [22] (**Chapter 2**). The porphyrin carbohydrate-based bioconjugates were obtained in 5–16% yield *via* condensation reaction of a carbohydrate aldehyde with pyrrole (**Scheme 5.6b**).

As previously reported in this chapter, the bioconjugation of the porphyrin derivatives with carbohydrates modulates their targeting, solubility,

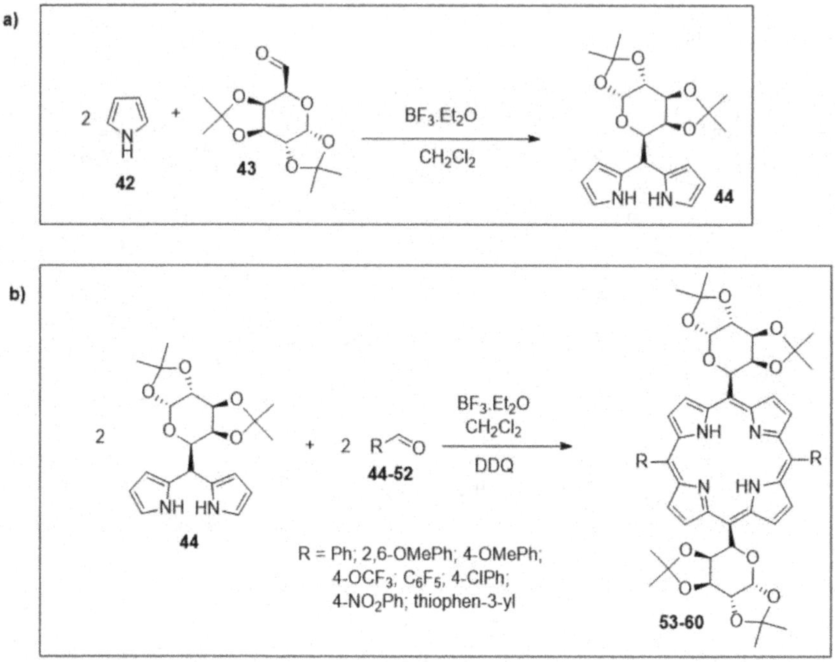

SCHEME 5.6 Synthesis of carbohydrate-based bioconjugate with porphyrin via condensation reaction of a sugar aldehyde with pyrrole producing *meso-bis-*glycosylated diarylporphyrins.

and uptake, improving their pharmacological properties. Moreover, this strategy may improve the physiological stability of porphyrin derivatives (*in vivo* studies) and reduce their toxicity.

REFERENCES

[1] Kurzyna-Szklarek, M., Cybulska, J., Zdunek, A., *Food Chem.*, **2022**, *394*, 133466.

[2] Pohl, N. L. B., *Chem. Rev.*, **2018**, *118*, 7865.

[3] Hall, M. B., Mertens, D. R., *J. Dairy Sci.*, **2017**, *100*, 10078.

[4] Navarro, D., Abelilla, J. J., Stein, H. H., *J. Ani. Sci. Biotechnol.*, **2019**, *10*, 39.

[5] Campos, V., Tappy, L., Bally, L., Sievenpiper, J. L., Le, K. A., *J. Nutr.*, **2022**, *152*, 1200.

[6] Bovin, N. V., Neoglycoconjugates: Trade and Art, in *Glycogenomics: The Impact of Genomics and Informatics on Glycobiology*, edited by Drickamer, K., Dell, A., Portland Press, **2002**, 143.

[7] Oppenheimer, S. B., Alvarez, M., Nnoli, J., *Acta Histochem.*, **2008**, *110*, 6.

[8] Liu, H. L., Bolleddula, J., Nichols, A., Tang, L., Zhao, Z. Y., Prakash, C., *Drug Metab. Rev.*, **2020**, *52*, 66.

[9] Weyant, K. B., Mills, D. C., DeLisa, M. P., *Curr. Opin. Chem. Eng.*, **2018**, *19*, 77.

[10] Khatun, F., Toth, I., Stephenson, R. J., *Adv. Drug Deliv. Rev.*, **2020**, *165–166*, 117.

[11] Wang, J., Zhang, Y., Lu1, Q., Xing, D., Zhang, R., *Front. Pharmacol.*, **2021**, *12*, 756724.

[12] Cermeno, M., Felix, M., Connolly, A., Brennan, E., Coffey, B., Ryan, E., FitzGerald, R. J., *Food Hydrocoll.*, **2019**, *88*, 170.

[13] Tornoe, C. W., Christensen, C., Meldal, M., *J. Org. Chem.*, **2002**, *67*, 3057.

[14] Rosa, M., Jedryka, N., Skorupska, S., Grabowska-Jadach, I., Malinowski, M., *Int. J. Mol. Sci.*, **2022**, 23.

[15] Lemieux, R. U., Bundle, D. R., Baker, D. A., *J. Am. Chem. Soc.*, **1975**, *7*, 4076.

[16] Roy, R., Tropper, F. D., Romanowska, A., Letellier, M., Cousineau, L., Meunier, S. J., Boratynski, J., *Glycoconj. J.*, **1991**, *8*, 75.

[17] Dixon, C. F., Nottingham, A. N., Lozano, A. F., Alexander Sizemore, J., Russell, L. A., Valiton, C., Newell, K. L., Babin, D., Bridges, W. T., Parris, M. R., Shchirov, D. V., Snyder, N. L., Ruppel, J. V., *RSC Adv.*, **2021**, *11*, 7037.

[18] Hirohara, S., Kawasaki, Y., Funasako, R., Yasui, N., Totani, M., Aitomo, H., Yuasa, J., Kawai, T., Oka, C., Kawaichi, M., Obata, M., Tanihara, M., *Bioconjug. Chem.*, **2012**, *23*, 1881.

[19] Stasio, B. D., Frochot, C., Dumas, D., Even, P., Zwier, J., Müller, A., Didelon, J., Guillemin, F., Viriot, M.-L., Barberi-Heyob, M., *Eur. J. Med. Chem.*, **2005**, *40*, 1111.

[20] Asayama, S., Mizushima, K., Nagaoka, S., Kawakami, H., *Bioconjug. Chem.*, **2004**, *15*, 1360.

[21] Vedachalam, S., Choi, B. H., Pasunooti, K. K., Ching, K. M., Lee, K., Yoon, H. S., Liu, X. W., *Med. Chem. Commun.*, **2011**, *2*, 371.

[22] Lindsey, J. S., Hsu, H. C., Schreimen, I. C., *Tetrahedron Lett.*, **1986**, *27*, 4969–4970.

CHAPTER 6

Porphyrin–Peptide Bioconjugates

6.1 INTRODUCTION

The use of peptide-based therapeutics has expanded exponentially since the first use of insulin to treat diabetes, discovered by *Banting* [1]. Nowadays, there are around 80 peptide drugs in clinical use, around 150 in clinical development, and more than 500 undergoing preclinical studies [2,3,4]. It is well-recognized that synthetic and natural peptide-based drugs show good tissue penetration and high selectivity and specificity for the selected target [5,6,7,8]. As selected example, Detectnet™, a ^{64}Cu-DOTA linked to somatostatin receptor antagonist peptide, was approved by FDA (US Food and Drug Administration) in 2020 as radiolabeled probe for positron emission tomography (PET) [9,10]. However, to the best of our knowledge, until 2023, there have been no peptide–porphyrin bioconjugates in clinical use, although there is a growing interest in their scientific development [11,12,13].

In this chapter, selected examples of peptide–porphyrin bioconjugates will be described, with particular emphasis on their synthetic methodologies, reaction mechanisms, and applications. It is not the aim of this chapter to present and discuss issues related to peptide synthesis. For further reading, additional literature is given [14,15,16,17].

6.2 SYNTHESIS OF PORPHYRIN–PEPTIDE BIOCONJUGATES

The bioconjugation of peptides with porphyrins is a very delicate process, prior to which several critical issues must be considered. The presence

DOI: 10.1201/9781003119265-6

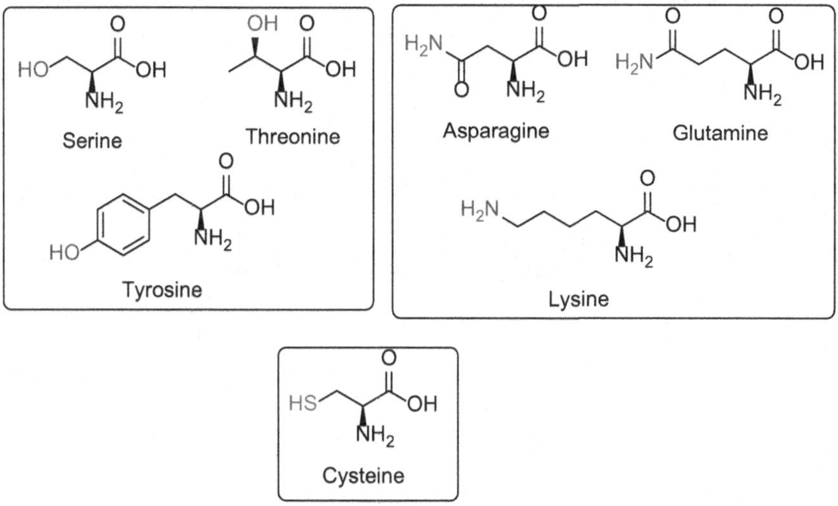

FIGURE 6.1 Amino acids typically used for bioconjugation.

of charges after bioconjugation can affect the function of the peptide sequence and consequently alter the interaction with the porphyrin. Moreover, the presence within the peptide sequences of different groups with similar reactivities demands very close control of the regioselectivity in order to avoid undesirable multiple conjugations. **Figure 6.1** shows the typical amino acids used for bioconjugation. All of them have nucleophilic groups that can give a mixture of products when bioconjugation is performed, if the reaction conditions are not strictly selected [11].

Nowadays, in order to obtain porphyrin–peptide bioconjugates with high selectivity and specificity, several methods have been reported, based on the functional groups involved in the conjugation [11,13]. **Figure 6.2** presents the three main conjugation strategies that are more often employed.

The Staudinger reaction [18,19] is based on the formation of a stable amide bond by reacting an azide with a phosphine molecule using mild conditions. *Umezawa* [20] described the synthesis of porphyrin–amino acid bioconjugates using this methodology (**Scheme 6.1**). 5,10,15,20-tetrakis(3-azidophenyl)porphyrin **P8** and the *N*-acetyl amino acid phosphinothioesters **1–3** were dissolved in anhydrous NMP (*N*-methylpyrrolidone) or dry DMF and reacted at 50°C for 20 h. Subsequently, after the addition of a small amount of H_2O, the reaction was left with stirring at room temperature for further 3 h. The obtained products were purified through

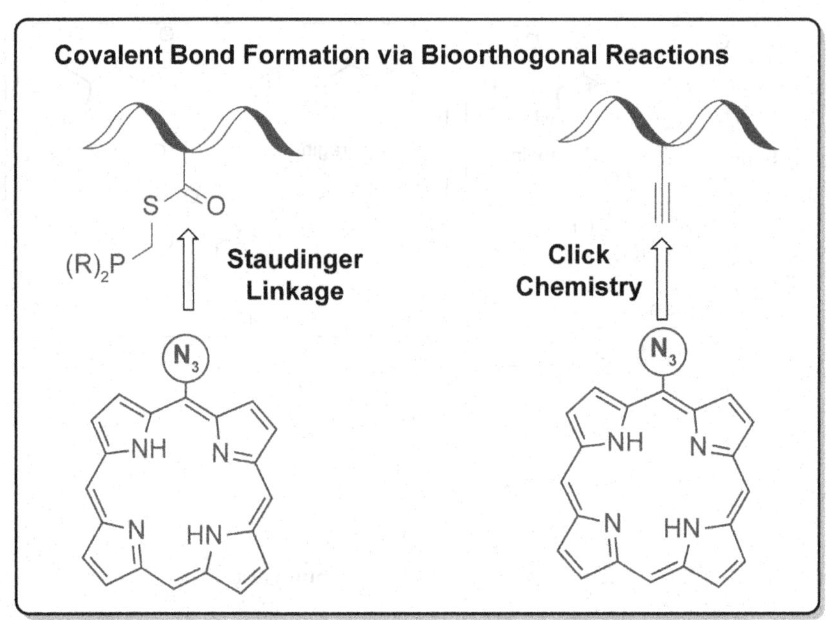

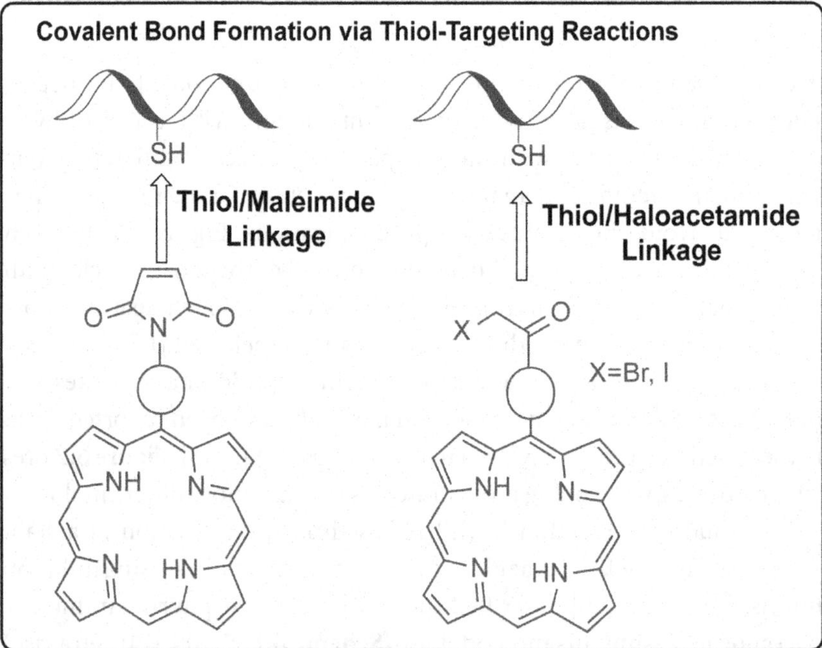

FIGURE 6.2 Peptide–porphyrin bioconjugation strategies often used in literature.

Source: Adapted from [11].

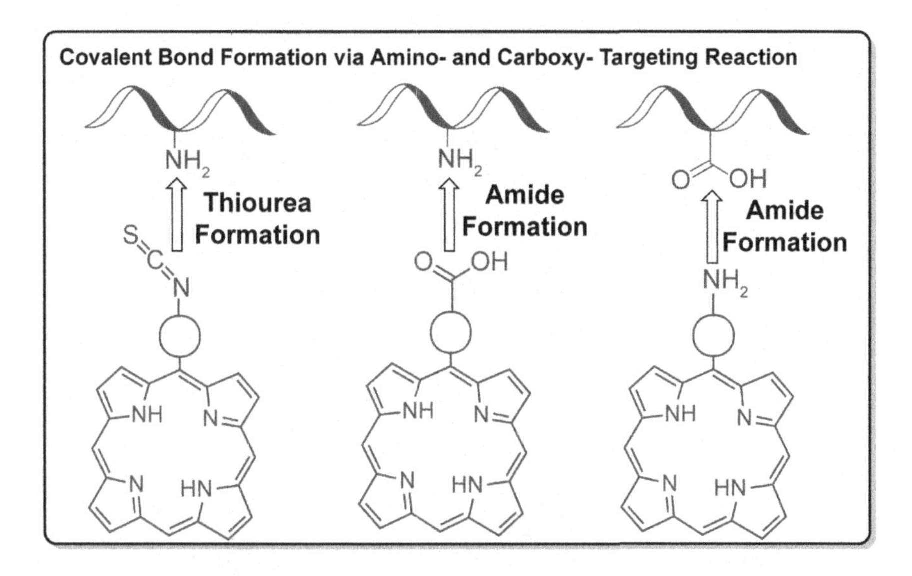

FIGURE 6.2 (Continued)

an alumina column chromatography, yielding the porphyrin–amino acid bioconjugates **4–6** in 71–90% yields. The accepted mechanism [19,21,22,23,24,25,26] for this reaction involves the nucleophilic attack of the phosphorous atom of phosphine derivative to the porphyrin's azide moiety, resulting in the intermediate **I**. This intermediate undergoes an intramolecular cyclization, forming a four-membered ring transition state **II**, which is then transformed into intermediate **III**, after the loss of N_2. This intermediate undergoes intramolecular cyclization, and the obtained tetrahedral intermediate **IV** will provide the amidophosphonium salt **V** that is hydrolyzed to form the amide **VI** and phosphine oxide **VII**.

Ghiladi [27] reported the synthesis of porphyrin–peptide bioconjugates using click chemistry as conjugation strategy (see mechanism in **Chapter 3**). As presented in **Scheme 6.2**, the authors performed the reaction of **peptide A** or **peptide B** with ethyne-substituted porphyrin **P18** using Cu(0) powder/copper(I) triflate toluene complex $[Cu(OTf)]_2$.tol as catalyst, DMF as solvent, and DIPEA as base, at 65°C, for 36 h (**Scheme 6.2**). The resulting protected bioconjugates were purified by HPLC, after which the amino acid residues of the peptide side chain were deprotected by adding a mixture of TFA (trifluoroacetic acid) (95%): H_2O:TIS (triisopropylsilane). Finally, diethyl ether was added to induce precipitation, and

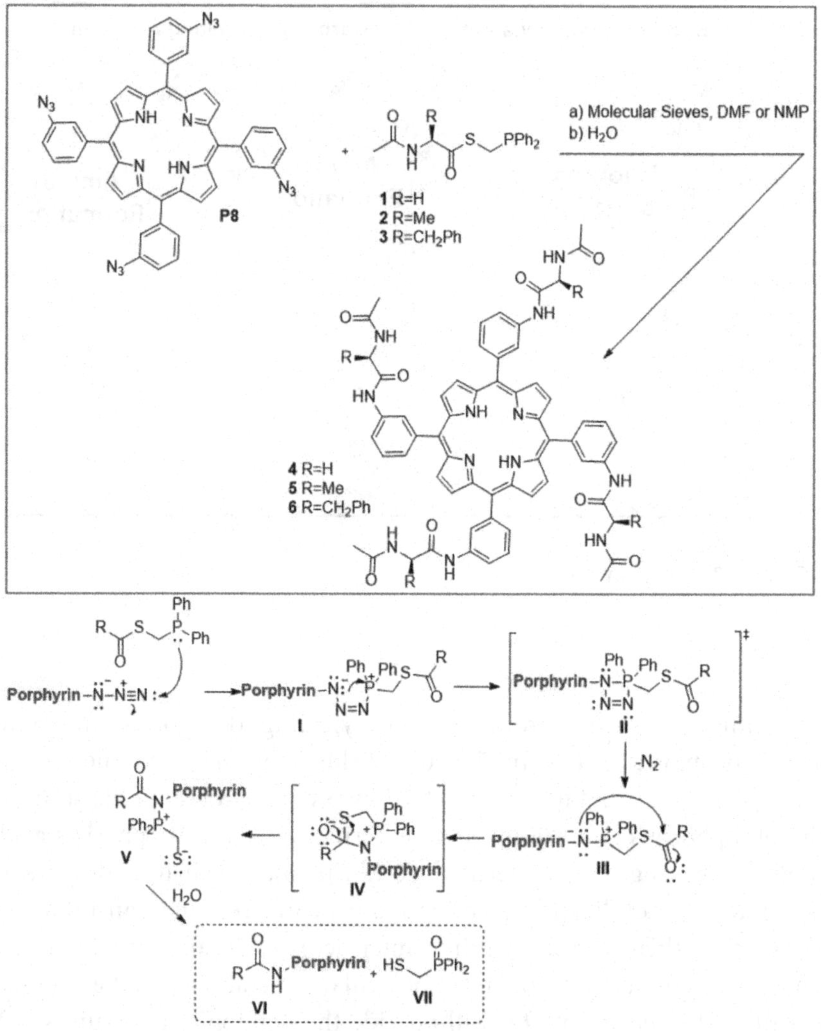

SCHEME 6.1 Synthesis of porphyrin–amino acid conjugates **5–7** and traceless Staudinger linkage mechanism.

Source: Adapted from [19,26].

the porphyrin–peptide bioconjugates were purified using semi-preparative HPLC, resulting in pure compounds **7–11** in up to 74% yield. These porphyrin–peptide bioconjugates were tested as photosensitizers for *in vitro* photodynamic inactivation of *Mycobacterium smegmatis*, showing bacterial inactivation of 3–4 log units, for concentrations between 0.25 and 0.5μM.

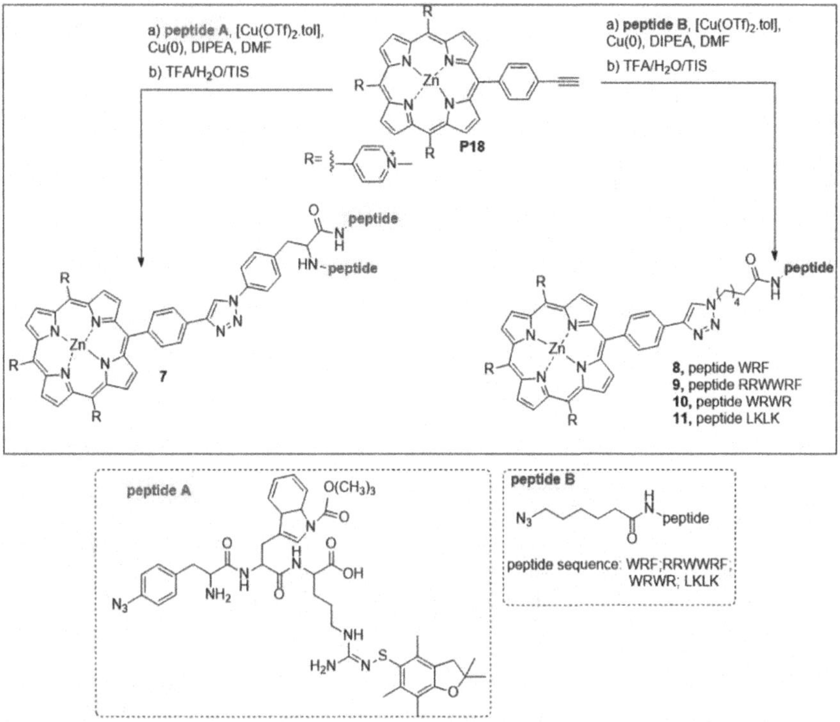

SCHEME 6.2 Synthesis of porphyrin–peptide bioconjugates **7–11**.

One of the biorthogonal strategies that is also being used for porphyrin–peptide bioconjugation involves coupling *via* SPAAC (strain-promoted azide–alkyne cycloaddition). The *Eggleston* group [28] used this SPAAC synthetic methodology to obtain a group of porphyrin-cell-penetrating peptide bioconjugates. Initially, the authors performed mono-amination of **TPP** using the methodology proposed by *Luguya* [29] described in **Chapter 2** and then performed acylation with dibenzocyclooctyne acid using EDC/HOBt activation, giving porphyrin **13** in 70% (**Scheme 6.3a**). The coupling of the azido peptides **14–17** with porphyrin **13** was carried out in DMSO and using pyridine as base (see mechanism in **Chapter 3**). The reaction was left in the dark overnight, at 25°C, (**Scheme 6.3b**). For purification, the desired bioconjugates were diluted with 1% aq. TFA and purified using a semi-preparative HPLC. Porphyrin–peptide bioconjugates **18–21** were isolated in 60–74% yields. The authors determined the cellular uptake and endosomal localization of the obtained molecules by confocal microscopy, in MC28 rat fibrosarcoma cells, and observed

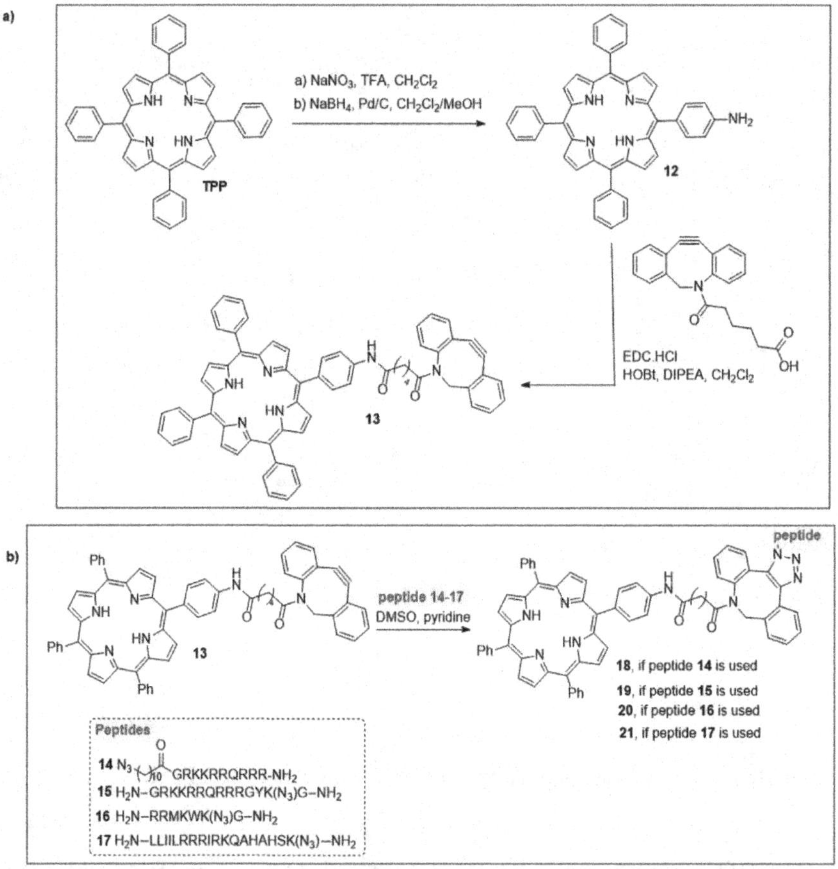

SCHEME 6.3 (a) Synthesis of porphyrin **13**; (b) synthesis of porphyrin–peptide bioconjugates **18–21** via SPAAC.

enhanced cellular uptake with the bioconjugates when compared with the non-conjugated porphyrin. The phototoxicity of bioconjugate **18** was also determined in MC28 human breast cancer, which showed that the porphyrin concentration required to induce 50% toxicity (LD_{50}) after illumination with blue light was ~40 nM.

As previously mentioned, cysteine is one of the most used amino acids for bioconjugation with porphyrins. Typically, the reactions involve the formation of thiol/maleimide or thiol/haloacetamide linkages. In this regard, *Xing* [30] described the synthesis of monomeric and dimeric porphyrin–peptide (**Scheme 6.4**) bioconjugates and evaluated its potential for effective photoinactivation and intracellular imaging of Gram-negative

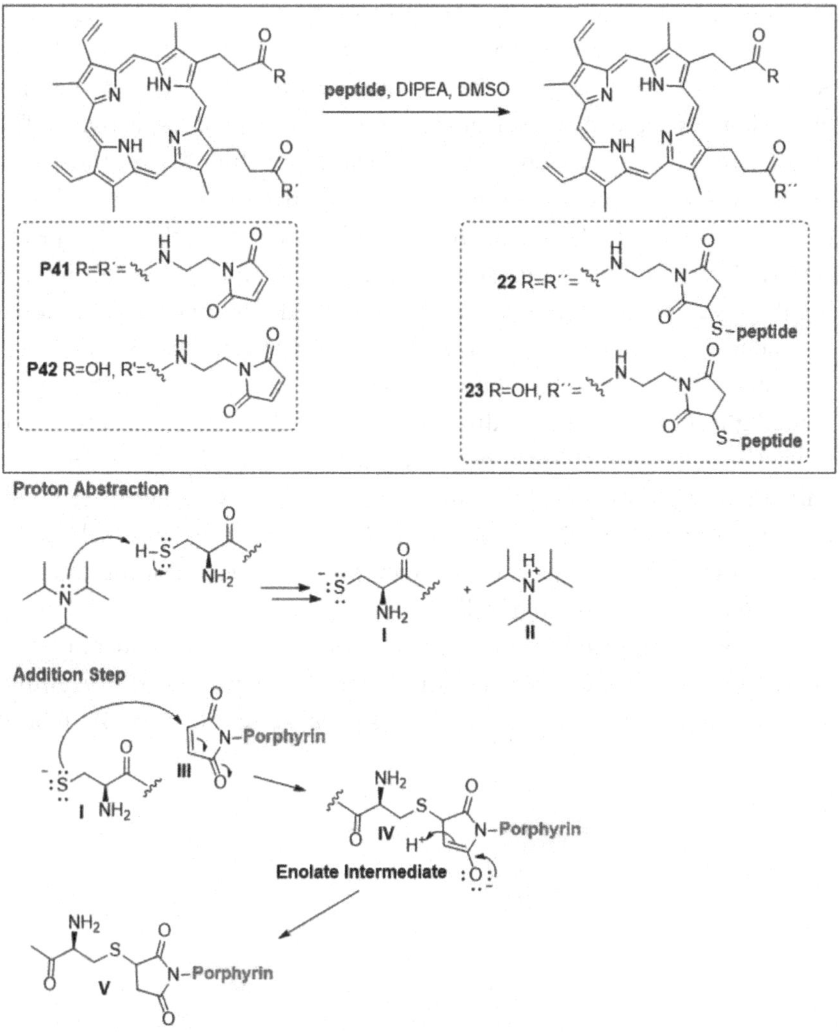

SCHEME 6.4 Synthesis of mono- and dimeric porphyrin–peptide bioconjugates **22** and **23** and base-catalyzed thiol-Michael addition.

Source: Adapted from [30,31].

bacteria strains. Initially, **protoporphyrin IX** was functionalized (see **Chapter 2** for further details) with maleic anhydride to introduce one **P42** or two **P41** maleimide moieties. Then, coupling of **P41** and **P42** to the cysteine-containing lipopolysaccharide-neutralizing peptide sequence YI13WF was carried out by dissolving **P41** or **P42** and the peptide in a solution of DMSO containing 10% DIPEA. Then, the products were precipitated with diethyl ether and purified using semi-preparative HPLC. After lyophilization, dimeric **22** and monomeric **23** porphyrin–peptide bioconjugates were obtained at 56% and 48% yield, respectively. The reaction mechanism involves the base-catalyzed thiol-Michael addition [31,32], where the reaction is initiated by thiol proton abstraction, followed by reaction of the highly nucleophilic thiolate anion **I** with the conjugate acid **II**. This anion will attack the π bond of maleimide **III**, resulting in a highly basic enolate intermediate **IV** that will deprotonate an additional equivalent of thiol, giving the intermediate **V** and another thiolate intermediate that continues the catalytic cycle. The authors observed that, upon white light illumination, the dimeric conjugate **22** presented higher photoinactivation against Gram-negative strains (*E. coli DH5a*, *S. enterica*, antibiotic-resistant *E. coli BL21*, and *K. pneumoniae*) when compared with monomeric conjugate **23**. In addition, **22** also presents some selectivity in recognizing bacterial strains over mammalian cells, leading to lower phototoxicity in mammalian cells.

Vicente [33] reported the synthesis of porphyrin–peptide bioconjugates *via* thiourea or amide bond formation. For the coupling with the amino acids or peptides, two different strategies were used (**Scheme 6.5**). In both

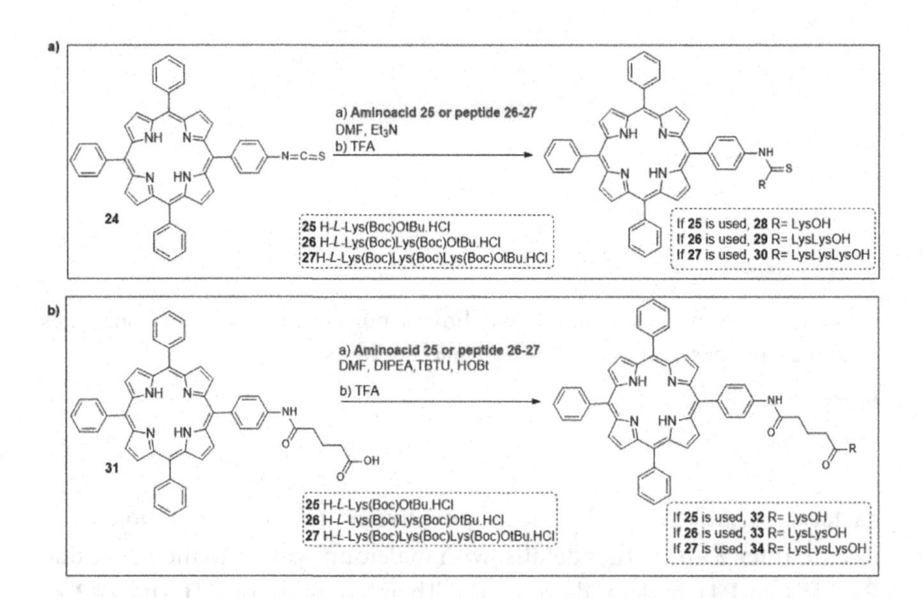

SCHEME 6.5 Synthesis of porphyrin–peptide bioconjugates **28–30** (panel **a:** thiourea covalent bonding) and **32–34** (panel **b:** amide covalent bonding).

cases, the authors used porphyrin **12** as starting material and converted the amino group into isothiocyanate by reaction with 1,1′-thiocarbonyldi-(2*H*)-pyridone, giving **24** in 84% yield, or into a carboxyl group *via* reaction with glutaric anhydride, giving **31** in quantitative yield. The coupling of porphyrin **24** to the amino acid peptide residue was performed in DMF, under argon, using Et_3N as base. The purification of the obtained products was performed using flash chromatography with silica gel as stationary phase. Finally, the deprotection of amino acid residues was performed using TFA, giving porphyrin–peptide **28–30** bioconjugates at almost quantitative yield. In turn, the coupling of porphyrin **31** with the amino acids or peptides was performed using HOBt/TBTU (1-hydroxybenzotriazole/2-(1-H-benzotriazole-1-yl)-1,1,3,3-tetramethylaminium tetrafluoroborate) as coupling agents. The porphyrin–amino acid or peptide bioconjugates were purified, as described previously. After deprotection, porphyrin bioconjugates **32–34** were obtained in quantitative yield. Comparing both strategies, the authors concluded that the latter provided higher yields of protected porphyrin bioconjugates, and that **32–34** presented adequate water solubility. All synthesized bioconjugates presented low dark cytotoxicity and were specifically observed in cellular vesicles and lysosomes.

An alternative strategy for the synthesis of porphyrin–peptide bioconjugates is based on solid-phase peptide synthesis (SPPS) [34,35] described by Merrifield (for which he was awarded the Nobel Prize in Chemistry in 1984) [36]. This method, typically used for the synthesis of peptides, is advantageous since the excess of reagents and products can be easily removed, the intermediates do not require isolation or characterization, and fewer protection/deprotection stages are required [37].

In 2022, *Wong* [38] described the use of SPPS in the synthesis of multivalent porphyrin–peptide bioconjugates. **Scheme 6.6** shows one of the structures synthetized by the authors. To a peptide attached to a rink amide resin **35**, 2-(4-formylphenoxy) acetic acid **36** was added, and the coupling reaction was performed using benzotriazol-1-yloxytripyrrolidinophosphonium hexafluorophosphate (PyBOP), DIPEA as base, and DMF as solvent. After 8 h, the obtained aldehyde **37** was washed with DMF, flushed three times with CH_2Cl_2, and fully dried by airflow. The condensation step was performed by mixing di(1*H*-pyrrol-2-yl-methane) with **37** in the presence of TFA and using CH_2Cl_2 as solvent. Then, the oxidation of the porphyrinogen to the corresponding porphyrin was carried out with *p*-chloranil in NMP (*N*-methylpirrolidone). The last step was the cleavage of **39** and deprotection of the amino acid residue using a mixture of TFA/

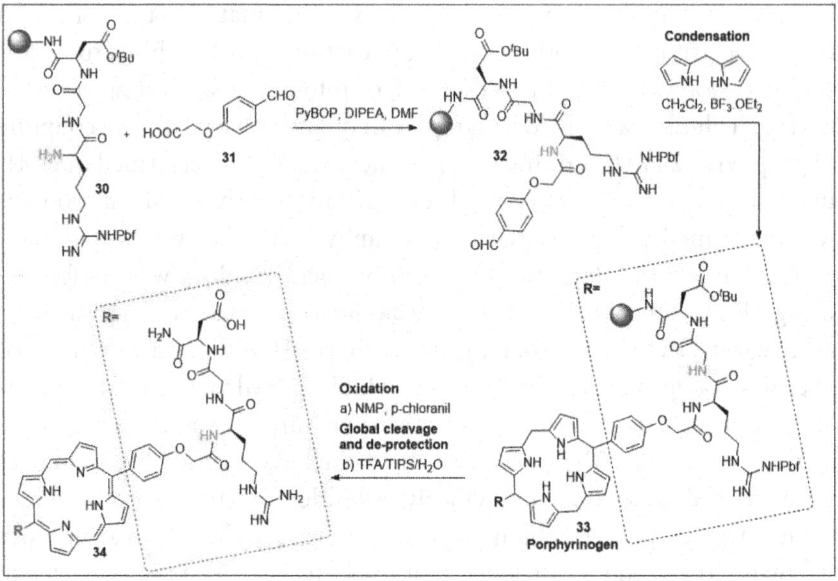

SCHEME 6.6 Synthesis of porphyrin–peptide bioconjugate **39** using SPPS.

TIPS/H_2O (95/2.5/2.5). HPLC analysis showed that the obtained product comprised 93% porphyrin **39**. The authors also tested CH_2Cl_2 as solvent and 2,3-dichloro-5,6-dicyano-1,4-benzoquinone (DDQ) or *p*-chloranil as oxidants. They concluded that N-methyl-2-pyrrolidone (NMP) was a better solvent and *p*-chloranil was easier to remove from the reaction medium in the work-up step. It should also be noted that this porphyrin–peptide bioconjugate did not show any relevant cytotoxicity against HeLa cell line.

The bioconjugation of porphyrins with peptides has proven to be an efficient strategy for preparation of molecules that combine the characteristics of both individual chemical entities. In sum, we should highlight that covalent peptide–porphyrin bioconjugates are, in general, biocompatible and show high specificity for targeting.

REFERENCES

[1] Vecchio, I., Tornali, C., Bragazzi, N. L., Martini, M., *Front. Endocrinol.*, **2018**, *9*, 613.
[2] Muttenthaler, M., King, G. F., Adams, D. J., Alewood, P. F., *Nat. Rev. Drug Discov.*, **2021**, *20*, 309.
[3] Fosgerau, K., Hoffmann, T., *Drug Discov. Today*, **2014**, *20*, 122.
[4] Lau, J. L., Dunn, M. K., *Bioorg. Med. Chem.*, **2018**, *26*, 2700.
[5] Sachdeva, S., *Int. J. Pept. Res. Ther.*, **2017**, *23*, 49.

[6] Jackson, J. A., Hungnes, I. N., Ma, M. T., Rivas, C., *Bioconjug. Chem.*, **2020**, *31*, 483.
[7] Meier-Menches, S. M., Casini, A., *Bioconjug. Chem.*, **2020**, *31*, 1279.
[8] Yang, S.-B., Banik, N., Han, B., Lee, D.-N., Park, J., *Pharmaceutics*, **2022**, *14*, 1378.
[9] Musaimi, O. A., Shaer, D. A., Albericio, F., de la Torre, B. G., *Pharmaceuticals*, **2021**, *14*, 145.
[10] Gutfilen, B., Souza, S. A. L., Valentini, G., *Drug. Des. Devel. Ther.*, **2018**, *12*, 3235.
[11] Giuntini, F., Alonso, C. M. A., Boyle, R. W., *Photochem. Photobiol. Sci.*, **2011**, *10*, 759.
[12] Biscaglia, F., Gobbo, M., *Pept. Sci.*, **2018**, *110*, e24038.
[13] Pathak, P., Zarandi, M. A., Zhou, X., Jayawickramarajah, J., *Front. Chem.*, **2021**, *9*, 764137.
[14] Ferrazzano, L., Catani, M., Cavazzini, A., Martelli, G., Corbisiero, D., Cantelmi, P., Fantoni, T., Mattellone, A., De Luca, C., Felletti, S., Cabri, W., Tolomelli, A., *Green Chem.*, **2022**, *24*, 975.
[15] Sharma, A., Kumar, A., de la Torre, B. G., Albericio, F., *Chem. Rev.*, **2022**, *122*, 13516.
[16] Hermanson, G. T., *Bioconjugate Techniques*, 2nd Edition, Elsevier, **2008**.
[17] Gutte, B., *Peptides: Synthesis, Structures and Applications*, Academic Press, **1995**.
[18] Bednarek, C., Wehl, I., Jung, N., Schepers, U., Brase, S., *Chem. Rev.*, **2020**, *120*, 4301.
[19] Van Berkel, S. S., Van Eldijk, M. B., Van Hest, J. C. M., *Angew. Chem. Int. Ed.*, **2011**, *50*, 8806.
[20] Umezawa, N., Matsumoto, N., Iwama, S., Kato, N., Higuchi, T., *Bioorg. Med. Chem.*, **2010**, 6340.
[21] Gololobov, Y. G., Zhmurova, L. N., Kasukhin, L. F., *Tetrahedron*, **1981**, *37*, 437.
[22] Gololobov, Y. G., Kasukhin, L. F., *Tetrahedron*, **1992**, *48*, 1353.
[23] Tian, W. Q., Wang, Y. A., *J. Chem. Theory Comput.*, **2005**, *3*, 353.
[24] Widauer, C., Grützmacher, H., Shevchenko, I., Gramlich, V., *Eur. J. Inorg. Chem.*, **1999**, 1659.
[25] Leffler, J. E., Temple, R. D., *J. Am. Chem. Soc.*, **1967**, *89*, 5235.
[26] Soellner, M. B., Nilsson, B. L., Raines, R. T., *J. Am. Chem. Soc.*, **2006**, *128*, 8820.
[27] Feese, E., Gracz, H. S., Boyle, P. D., Ghiladi, R. A., *J. Porphyr. Phthalocyanines*, **2019**, *23*, 1414.
[28] Dondi, R., Yaghini, E., Tewari, K. M., Wang, L., Giuntini, F., Loizidou, M., MacRobert, A. J., Eggleston, I. M., *Org. Biomol. Chem.*, **2016**, *14*, 11488.
[29] Luguya, R., Jaquinod, L., Fronczek, F. R., Vicente, M. G. H., Smith, K. M., *Tetrahedron*, **2004**, *60*, 2757.
[30] Liu, F., Ni, A. S. Y., Lim, Y., Mohanram, H., Bhattacharjya, S., Xing, B., *Bioconjug. Chem.*, **2012**, *23*, 1639.
[31] Northrop, B. H., Frayne, S. H., Choudhary, U., *Polym. Chem.*, **2015**, *6*, 3415.
[32] Nair, D. P., Podgorski, M., Chatani, S., Gong, T., Xi, W., Fenoli, C. R., Bowman, C. N., *Chem. Mater.*, **2014**, *26*, 724.

[33] Sibrian-Vazquez, M., Jensen, T. J., Fronczek, F. R., Hammer, R. P., Vicente, M. G. H., *Bioconjug. Chem.*, **2005**, *16*, 852.
[34] Meier-Menches, S., Casini, A., *Bioconjug. Chem.*, **2020**, *31*, 1279–1288.
[35] Gómez, J., Sierra, S., Ojeda, C., Thavalingam, S., Miller, R., Guzmán, F., Metzler-Nolte, N., *JBIC*, **2021**, *26*, 599.
[36] Merrifield, R. B., *Angew. Chem. Int. Ed.*, **1985**, *24*, 799.
[37] Petrou, C., Sarigiannis, Y., Peptide Synthesis: Methods, Trends, and Challenges, in *Peptide Applications in Biomedicine, Biotechnology and Bioengineering*, edited by Koutsopoulos, S., Woodhead Publishing—Elsevier, **2018**.
[38] Wu, Y., Chau, H.-F., Yeung, Y.-H., Thor, W., Kai, H. Y., Chan, W.-L., Wong, K.-L., *Angew. Chem. Int. Ed.*, **2022**, *61*, e202207532.

Index

Note: Numbers in *italics* indicate figures and numbers in **bold** indicate tables on the corresponding page.

A

B

C

R

S

T

Printed in the United States
by Baker & Taylor Publisher Services